THE VOICE THAT PRECEDES THOUGHT

Other Books by Tara Singh

COMMENTARIES ON A COURSE IN MIRACLES

"LOVE HOLDS NO GRIEVANCES"
THE ENDING OF ATTACK

A COURSE IN MIRACLES
A GIFT FOR ALL MANKIND

THE FUTURE OF MANKIND
THE BRANCHING OF THE ROAD

HOW TO LEARN FROM A COURSE IN MIRACLES

THE FORCES THAT SWAY HUMANITY

FROM THE CITY TO THE VEDIC AGE OF SHINING BEINGS
EXTENDING THE WILL OF GOD

MYSTIC AND OTHER POETS

THE VOICE THAT PRECEDES THOUGHT

TARA SINGH

FOUNDATION FOR LIFE ACTION
Los Angeles

The following portions of this book were originally published separately by the Foundation for Life Action. Each work has been completely revised. An * indicates the work has been expanded as well as revised for this edition. An † indicates the work was originally issued as one of the pamphlets in the *Holding Hands With You* series. *Silence* has not been previously published.

Gifts From The Retreat (12/80); **Gratefulness* (4/81); **Productivity Eliminates Unfulfillment [Productivity* in this edition](4/81); †*Holding Hands With You* (approx.4/81); **Time And The Timeless: The Relationship Between The Known And The Unknown [Time And The Timeless* in this edition](5/81); †*The Preparation* (6/81); *Excerpts From The Forty Days In The Wilderness* (10/81); †*Profile Of Our Age* (11/81); †*Non-Commercialized Life/The Path Of Virtue* (3/82); *Invitation To The One Year Non-Commercialized Retreat* (5/82); †*"I Am In Need Of Nothing But The Truth"* (8/82); **The Voice That Precedes Thought [The Voice* in this edition] (9/82); †*The One Year Non-Commercialized Retreat* (3/83).

Library of Congress Cataloging in Publication Data
Singh, Tara, 1919-
The voice that precedes thought.
Contents: Gratefulness — Productivity — Time and the timeless— [etc.]
1. Gratitude. 2. Time. 3. Silence.
4. Spiritual life. I. Title.
BJ1533.G8S565 1985 179'.9 85-27573
ISBN 1-55531-002-8 Limited edition, hardbound
ISBN 1-55531-003-6 Softcover

The material from *A Course In Miracles* and *The Gifts Of God* is used by permission of the copyright owner, the Foundation for Inner Peace, Tiburon, California. Cover illustration by William Blake: *God Setting His Compass On The World.* Back cover photograph by Tey Roberts.

Dedicated to Lucille Frappier and Jim Cheatham for their long hours of devoted work. Without their assistance, this book would not have been possible.

Acknowledgments

I am grateful to the following friends for their help in the preparation of this book: Connie Willcuts, Norah Ryan, Clio Dixon, Aliana Scurlock, Frank Nader, Kris Heagh, Charles Johnson, and Johanna Macdonald. And for those whose support has made it possible, my appreciation is also extended: John Williams, Mary Titus, Joann Nieto, Rachel Logel, Selina Scheer, Nancy Marsh, Richleigh Heagh, Sandra Lewis, Richard Michael, Ted Ward, Acacia Williams, Steve James, Lara Petiss, Jean Kohn, Don Leach, and Malcolm Lewis.

In addition, I am thankful to the following friends who were instrumental in preparing the original editions of many of the books which make up The Voice That Precedes Thought: *Joy Kealey, Nolyn Johnson, Eileen Callahan, Pamela Kawin, Merilee Forestal, Chris and Martha Stedman, and Acacia Downs.*

Contents

Preface

A simple thing such as knowing one's own function in life
has become obscure and difficult.
Yet in each human being there is an inner calling
and the abilities one is born with to implement it.
The extension of his inner calling
would naturally introduce man to his function
and relate him with the universal Laws of creation
of which he is a vital part.

It has taken me over fifty years to have some inkling
of my calling in life.
Yet the function is ever there to be heeded.
As awareness began to awaken me
to a state beyond the body senses,
effortlessly a sense of rightness changed my values
by relating me to Life Itself.

What a wonder to discover that inherent in the function
is the freedom from the limitation of body senses!
To be related to Life Itself
rather than to thoughts *about* Life
would transform the very foundation
of our present educational system.

The discovery of one's own inner calling
is an entirely different thought system
for it does not acknowledge separation
nor the external as real.
The peace and beauty of one's eternal Self
cannot be regulated nor pressured by external thoughts.

In present society where education is mandatory,

the child is schooled and removed further and further
from insight and from the light of his own awareness.

He is made to learn geography, history,
national prejudices and concepts,
in spite of his innate resistance
to what is false and abstract.

The child is coerced and pampered and made to comply.
What he is taught is spelling, grammar and mathematics.
But what he learns is fear, conformity,
and the power that authority has over him.
It is the beginning of internal conflict and violence.
He learns to ignore himself
as he is taught limitation, insecurity, and "might is right."

Having acquired a skill for survival,
he usually accepts being a wage-earner
and hence remains a mercenary.
The routine allows no space
nor Divine leisure to know anything other
than his nationality and acquired profession.
The need for outlets becomes all-consuming.
Routine work and outlets go hand and hand.

In this accelerated age of unfulfillment,
the need for outlets supports a trillion dollar industry
next only to the military in expenditure.
Without this constant waste and abuse
of human energy and of the planet's resources,
the economic system of the affluent world
might well collapse.

The child is prepared for a way of life
for which he is then held responsible.
He seldom realizes who he is as a part of creation,
nor his own Divine function as a human being.
As a human being,

he is the focus of planetary forces upon the earth
and brings to it a light, not of the sun,
but of his own eternal Self —
a light that never departs from the planet.*

Fortunately, I was spared the schooling
of relative knowledge
during my adolescent years.
But in my youth,
in the company of the wise of unpressured space,
I learned the ethics of humanism and simplicity.
This led me, in my later years
to make acquaintance with Lao Tzu
Nanak, Socrates, Thoreau, Emerson,
and other men and women of True Knowledge.

Even then, it took years to outgrow and undo
the values of self-centeredness the world imposed.
The human brain is so impressionable
it involuntarily gets contaminated.
To unlearn is basic but arduous
for a man of externalized existence.
There is not the leisure to free and to renew the brain
from tension and stimulation.

In my later life,
the flashes of insights were more frequent
and freed me from the thought system
of self-imposed limitation.
The call from within energized me
and had its own cleansing.
It was impersonal; hence, not emotional.

The newness made everything one knew irrelevant.

---

*Mr. Singh discusses in-depth the issues involved in the education and upbringing of a child in his book, *How To Raise A Child Of God*. (Editor)

It imparted the joy of being alive
WHENEVER I HAD SOMETHING TO GIVE.
Givingness became more intrinsic than wantings.
I learned that man is energized by the action of giving.

This transformation opened new vistas and an awareness
of forces at work beyond appearances.
Suddenly I realized
I was being educated all along by the Unknown
with a gratefulness that transcends time.
All through my life it inspired me
in the midst of distress.

If man could know the truth of gratefulness,
it would liberate him.
Gratefulness is everlasting.
It is not necessarily for a thing.
It is a contact with a state that is alive in you.

Having Something To Give freed me from a sense of lack
and a new lifestyle began to evolve.
The potential of those moments,
having become part of the One Mind,
seemed so vast it could transform the world.

Occasionally I felt a tremendous need
to be alone for a time
and make closer contact with the Unchangeable,
the wisdom of Silence.

After a period of solitary seclusion,
I became sensitive to a purer energy,
often with the wisdom of knowing what *not* to do.
It proved to be a period of unlearning
and had in it the power of non-action.

I was directed to do workshops in the New World.
To this I had resistance for my words were not yet honest.

Preface

I lacked the power of responsibility
and the power of the True Word.
I refused to interfere in other people's lives
with half-truths.
Thus, I would not teach what I had learned from another.

Time went by and I walked under the sky without direction.
But the uncertainty that agonizes the mind was absent.
It is a joy to have no plans of one's own
and to be with the freedom of uncertainty.
To pursue a projected goal, no matter how noble,
is invariably self-centered and has its consequences.

The intimation *insisted* that I do the workshops
and I was awakened to a clarity within —
ever impersonal — which directly understood:

> "Only by removing other people's obstacles
> to True Knowledge will yours be removed.
> Try it. The work will be blessed."

This book, *The Voice That Precedes Thought,*
is primarily comprised of sharings that emerged
out of weekend workshops, ten-day retreats,
and the Forty Days in the Wilderness.
These books summarize the period
prior to the action of non-commercialized life,
the One Year Retreat.

Gifts from the Retreat and *Gratefulness*
were the first books that emerged.
Productivity, Time and the Timeless and the others followed.
The *Holding Hands* were sent
to all workshop and retreat participants
as an extension of the one-to-one relationship
which was offered.

The *Voice* that precedes thought is not personal,
nor is there an "other" in it.

It dissolves the separation
and is accompanied by the awareness of Grace
that silences the mind.

After the Forty Days Retreat in 1981,
consistent with the song we sang,
"We come not to *learn* but to *BE,*"*
the period of workshops and retreats came to an end.
They had served as a preparation for the real work ahead:

THE NAME OF GOD CANNOT BE COMMERCIALIZED.

The advent of *A Course In Miracles*[1] upon the planet
brought me to wholeness,
often to a consistency at all levels of my being
to heed my function.
Life brought me into close contact with its scribe
and I read the Course with a reverent heart.
At times it imparted the space
that ended the interpretations.
What a blessing it is to heed
the eternal words of Absolute Knowledge.

The Course is here for all times and beyond time.
It will transform coming generations
with the light it imparts.

A Course In Miracles is the first scripture
ever to originate in the New World and in English.
It endows mankind with:

"By grace I live.
By grace I am released.
By grace I give.
By grace I will release."[2]

---

*This refers to the song, "The Seventy." See page 282.

Preface

It means exactly what it says.
What would it take to know the power of its truth?
It is a direct knowing that awakens True Knowledge.
It invokes the knowing of Divine Laws
of which man is an extension.

On Easter of 1983,
the One Year Non-Commercialized Retreat began
with a group of serious students
for the day-to-day study of *A Course In Miracles.*
No tuition was charged.

Out of this One Year Retreat emerges a new thought system
not limited to manmade concepts
and fragmented belief systems.
The direct action is consistent
with the vibrations of the New World
to come to new consciousness of:

"IN GOD WE TRUST."

The Foundation for Life Action has evolved as a school
for *Students Capable of Going Beyond Instruction.*
It offers the intimacy of the one-to-one relationship
and a lifestyle of:

Resist Not Evil.[3]

Love Ye One Another.[4]

Nothing Real Can Be Threatened.[5]

I am not a body. I am free.
For I am still as God created me.[6]

The Foundation for Life Action started on the principle
of the one-to-one relationship
and, for us, it is the human being that comes first.
We are not religious.
We are men and women who endeavor to lead a life of ethics
and live by Eternal Laws.

The Foundation for Life Action,
a federally-approved, non-profit, educational foundation
has emerged as a school
dedicated to non-commercialized life.
It does not accept charity, nor seek donations,
nor does it own a community.

Because the Name of God cannot be commercialized,
and ill-earned money begets other vices,
the School charges no tuition.
It has evolved its own integrity
of bringing the student to:

— Being Self-Reliant, Productive,
and Living a Non-Commercialized Life,

— Extending His Own Intrinsic Work
and Not Working For Another, and

— Having Something of His Own to Give
and Never Taking Advantage.

The work of the Foundation is to be consistent
with the Call to Wisdom of the Founding Fathers
and "In God We Trust."

Tara Singh

Introduction

There is a memory of God within us that can never be contaminated. Nothing can take it away. A person who has discovered his Reality is no longer regulated by time or by anything external. He has silenced the reactions within himself.

But one must meet the Timeless in the Present. The present moment is religious. If you deny the Present, you will also deny Love. And if you deny Love, you will deny your own peace.

To be motivated by fear is never to know peace, never to know the purity of your own being. And for the internal awakening, only you are responsible.

The most essential thing in life is silence. Inspiration brings one to silence for it changes one's state of being. The person who comes to this stillness has a blessing to impart. He is content and renewed by the Love of God — the Source of Life. He finds that HE is not outside of himself. HE is not a Name. HE just is.

A few moments of this is worth all of one's life.

* * *

The books that make up *The Voice That Precedes Thought* were written to help create an internal atmosphere of receptivity that can receive the Given — that which is beyond thought.

Create the space to read it with a quiet mind, for it is your stillness that becomes the fertile ground for truth to take root and grow.

BOOK ONE

Part I

Gratefulness

Gratefulness

Gratefulness is the silent peace of God
that surrounds us at all times.

There is a state of being where one is always thankful.
There are no means to it, for:

Freedom is at the beginning and not at the end.
J. Krishnamurti

Yet we could know this intellectually
and never realize that which is eternal about us.

It is the actuality that is valid,
the state itself — not the words about it.
Therefore, for those who are attentive
and ready for a new beginning,
there is something that words cannot communicate
but the spirit of gratitude can.

Once you are touched with that gratitude
you begin to discover it each day,
with every incident.
It is no longer a thing of memory.

Gratefulness brings you to an intensity of silence
where the duality of opposites end.
This thankfulness and gladness is something eternal.
You are thankful to all that sustains life.

Being thankful for what someone has given you

is human and nice
but it does not inspire you out of yourself.
You do not expand beyond your own limits.
It does not introduce you to your own boundlessness.

Can you see that without the planets,
without the vegetation,
without the brother,
you could not exist?
It ends all words — a gladness of gratitude without name.

Suddenly, unasked,
the veil is lifted
and you perceive the gratitude
of all things in nature
singing songs of their perfection.

The vastness of the ocean,
the stars in the sky
reveal to you your eternity.

Gratefulness is the key, the purity,
the freedom from all that is earthborn.
It imparts the essence of your Being.

The one who is grateful is of a still mind.
When you are grateful,
you want to spread your joy upon the planet.
Your expression is inspired by gratefulness;
it sustains you
for it is already with you.

Your life is blessed then
and your heart sings songs of adoration
of the perfection of God.

Gratefulness brings

the Kingdom of God to earth
and everything in creation awaits its blessing.

Gratefulness is a means of communication with the Creator.
It is all-encompassing.
There is only the today which becomes an eternal today
because it is not subject to day and night.

You could not then be subject to worries and anxieties.
You would never be busy
for you would always be present.

When you are present,
then the day is forever
because the present is forever.
The present is eternal.
Yesterday and tomorrow are not.
They are of time.

When you rejoice in being,
you are not aware of time.
And neither are you with words.

What then is gratefulness?
Gratefulness is a recognition of who you are.
Recognition is of your own identity and what is of God.

Gratefulness is your discovery of the perfection that is.
Gratefulness is not something you seek to become.
It is something you recognize the truth of.
It cannot fail.
Gratefulness makes everything possible.

We think gratefulness comes and goes,
but gratefulness is eternal.
Peace is eternal.
Love is eternal.

Because they are eternal,
they are not personal.
Because they are not personal,
they liberate you from the personal.

To move away from the personal
and touch upon the eternal
is to make right use of the brain.
Whenever the brain touches the eternal,
then it is grateful.
It is at peace.
Peace is of God.
How simple, yet how revolutionary it is.

When you are in a state of gratefulness,
nothing is external to you.
Gratefulness does not judge
because it sees the One Life.

Therefore, do not limit yourself —
see the boundlessness of gratefulness
that cannot be limited.
Man is caught in his own limitations.
Yet the limitations are all illusions.
The truth of man is that he is limitless
for he shares the Mind of God.
And the Mind of God is everywhere.

There is never a moment
when the blessing of God is not with you.

When you have gratefulness to impart —
love to impart —
then some change has taken place upon the planet.
The world need never be the same.

Gratefulness is the continual renewal

at the time level
of contact with the timeless.

Gratefulness introduces us to that one instant
in which the timebound personality meets eternity.
Wherever that takes place,
a miracle has happened.

There cannot be gratefulness without seeing your perfection
and the perfection of the whole.
Then there is only adoration.

Are you echoing that appreciation, that gladness?
Then you have the right relationship
with the universe,
with God.

You relate with the eternal,
rather than the limitations of personality.
And you have a responsibility
never to be part of deception.

Without gratefulness,
knowledge becomes a deception.
Thought, itself a deception, is afraid to touch Life.
This is a great discovery of man:
either thought rules you
or you rule thought.
When you rule thought
you are not of this world.

The stable purity of gratefulness
is the mighty strength
that formulates no external thoughts.
This is innocence.

Innocence is a space within one

that is pure and untouched by thought.
What is pure has no thought in it.

Once you know the purity of your own being,
you are not then seeking God or truth.
Thought disappears.
Even time ceases.
And that you can experience directly.

The listening within silences thought,
having received what is beyond thought.
If we could listen, then we come to silence.
Thought has a place,
for there is the precise and factual thought
that carries the silence with it.
It has the ability to silence.
And one is grateful and appreciative.

When you are truly grateful,
it may be the first time you have
an intimation of what the energy of love could be.
That is purity.

But where there is no gratefulness,
one is not touched by the words of truth,
nor by the silence the words contain.

The silence comes in when there is real interest.
And what would that interest be?
Real interest introduces us to gratefulness,
to an inspiration within.
When you touch upon this inspiration,
there is the joy of gratefulness.

In the true sense,
gratefulness is to Life.
This blossoming has to take place within.

The key is joy for what you receive from Life.
Then you are part of Life.

When you are grateful,
when you are in the state of purity
which is accessible to each person,
then you are not a body.

And the spirit within the body meets most gently
and lovingly the needs of the body.
It does not hoard or project,
and therefore it is liberated from time and consequences.
It lives a simple life
and becomes aware of the primary needs
from the Divine point of view.

Gratefulness does not know a lack.
It trusts in the Will of God
and leaves God's things to God.
It knows that
for what you are grateful you will never be denied.

All else is duality, fear, and selfishness,
bound to the body and its sensations.

The cause of all error is man's lack of trust
in the Will of God.
Thought does not go near it
because that is where thought will be dispelled.

We cannot rely on the brain,
or the limitation of thought,
but we can rely on the Will of God.

Is there such a thing as a crisis
to one whose life is not externalized?
We are talking about a state that precedes crisis,

that precedes thought.
We have to outgrow thought.

It is we who choose to give thought authority over us.
Could you question your own thought,
knowing that silence is superior?

You can confront thought with innocence.
But the greater wisdom is to acknowledge
that there is a relationship with Life.
Life supports any action that is of the NOW

If you are grateful for whatever the situation is,
then your thought is unemployed.
No circumstances can contain you.

The spirit is fearless
because it is not concerned with consequences.
It does not get into time or reaction,
not even into correction.
What holds you back are your own thoughts,
opinions, and attachments.

You cannot let the externals regulate you.
This means that you are to live by something
that is universal, that is timeless.
Your values change.

I would like to share with you my own experience.
Not too long ago,
something happened to me: I had a good rest.

In this relaxation
I saw what was happening to thought.
When the gap between the thought widens,
perception takes place, insight —
something involuntary.

The span of interest increases.

Insight is far wiser and swifter than thought.
From that state, I put a question to myself:

"Why do I need success?"

The instant I asked it,
immediately I was freed from it.
Do you know what it means to be free of seeking success?

You would have to thank each day a million times,
and you still could not do it justice.

It is essential for the human being
to learn what relaxation is
for gratefulness is a flower of relaxation,
as confusion is the outcome of stimulation.

The gift of relaxation causes
the unessential things of life
to fall away without effort.
It clears the way for gratefulness.

You begin to master circumstances.
External forces no longer rule you.
Sentimentality and intellectuality are over.
It has a completeness to it —
a sacredness to it.

Your entire relationship with people changes
because you do not relate with anyone
on the basis of motives.
You are free from the mania of advantages.

At this point,
my life became productive

because now I could be Real.

I do not pursue my projections, my knowings.
Now I can afford to be with you, totally,
at any moment we meet.

In this state,
You discover what it means
to be in harmony with AS IS
It is another discovery of gratefulness.
Something sacred and creative has brought you to that point
and now you are impeccable.

The discovery of AS IS,
which inspires the spirit of gratefulness,
is another action.
You become part of the Action of Life
rather than your own personal projections.

A still mind has no projects and no plans.
It is another relationship
with the reality that it is —
where Life is One
and all is Love.

Demand being in a state of stillness
that dissolves all questions and answers.

Thought and reason cannot do it
for inherent in thought is instability.
But the will can.
This is not the will of thought,
but the will which is a gift of God to man.
It does not fluctuate.
Its origin is Love.

Then you will see how gratefulness

begins to unfold in your life.
Probably then,
the real work begins.

It is not personal
and you are not afraid of uncertainty anymore,
for your eyes see beyond appearance.
They see that which is —
and God alone is.

God alone is,
and He surrounds you from head to toe
with His blessings.

You have become a part of Grace
which is beyond the words,
where the separation ends,
and seeking is no more.

What you do then is sacred and blessed.
It matters little what it is.
It is Divine Intelligence at work.
Nothing is outside of It.
You walk freely and purely,
untouched by anything.

And if it is His Will,
you help bring your brother to the perfection that he is,
never making him dependent.

A man of God is not a reformer.
He is an awakener.
He awakens you to your own reality.

Some time ago,
I opened the Bible at random
and read a passage about Christ at the Last Supper.

And he took bread,
and gave thanks,
and brake it . . .
Luke 22:19

It took me several months to ever again
read anything from the Bible,
so inspired was I by the gratefulness He expressed.
When you have really read it,
you have come to the actuality of that experience.

We read, but we never inherit the vitality
of the Scriptures, or the truth.

Once you have harnessed that energy,
it revitalizes you.
It puts you in that state.

Can you even imagine the gesture of breaking the bread?
What perfection it would have had!
And the offering of it —
every movement as if timeless,
out of stillness.

What a relationship of gratefulness with God,
of gratefulness with yourself.

This One lived it.

Real gratefulness would be the application.
One has to live it.

Gratefulness is born out of your eternity of being.
And that gladness you leave for all mankind —
for the whole planet — for eons to come.

If you are not with gratefulness,

then know that your potentials
are not being invoked.

What would it be but foolishness
not to recognize your own potential?
You walk with the richness of creation
when you know the gladness within.

The gladness that is not timebound has its own strength.
No mortal can stop its action.
That gratefulness has no opposites.

I am talking about something eternal,
something beyond personality.
"Love ye one another" (*John 13:34*) becomes possible
and you have something to give forever.
For truth can only be shared
when there is that gratefulness,
when there is that love of truth.

The light of truth dispells
what a hundred suns and a thousand moons cannot.
It is the light of truth that awakens.
How can you pass it by?

Gratefulness is inherent in truth.
It has its own perfection,
its own completeness.
It has its own wisdom, its own wholeness.

The love of your heart,
the gentleness of your being,
the gladness within
is changing what is on the planet.

Where gratefulness is there are no problems.
You discover that goodness is the Law of Creation.

Goodness is a necessity of Life.

What takes place is the recognition of the blessing,
the bliss, the completeness, the perfection.
And you are lost in it.

A grateful person does not want things.
He is an expression of simplicity and purity upon the earth.
He wants nothing from another,
thus he is independent.

> Gratefulness is ever alive and independent.
> It is the light that wants nothing.

To see the perfection of the God-created world
is to be grateful.

Seeing perfection will bring stillness.
Everything emerges out of stillness.
A green leaf out of a dry twig grows in silence,
and at every moment it is perfect.

Everything in creation is extending perfection.
And it exists for you, the holy Son of God.
When you see the perfection,
you are surrounded by it.

Would you not then be inspired out of pettiness,
that creation exists just for you?
Would you not fill the universe with your gratefulness?

Gratefulness is one's richness.
It wants nothing.
It has no preferences, no choices.
And therefore, it holds a kind view of human beings.

Yet man, who survives by creation,

destroys and exploits it.
Is that not ingratitude?

Ingratitude is self-destructive.
Where there is ingratitude,
there will never be love.

Let us say ingratitude is the obstacle.
Where there is ingratitude,
there is limitation because we are with the partial.

As a communist, one is limited.
As a capitalist, one is limited.
The minute you become limited,
your ingratitude towards the "enemy" or the "opposite"
is there too.
You do not have to read the newspapers
to see that it is inherent in nationalism, in you.

Limitation is the ingratitude,
thinking that you are not part of the Mind of God.
Know that when you step into ingratitude,
you are not consistent with who you really are.

Only ingratitude could resort
to projections, plans and activity.
Ingratitude is an activity of unfulfillment
that is always looking externally for its fulfillment.

When the obstacles are removed,
the illusions are removed,
then we see that you and I share the One Life, the One Mind.
It is our ingratitude towards the brother
that is the obstacle.
By limiting ourselves to self-interest,
we are deprived of the eternal.
We are made external to that which is whole.

The whole knows no separation.
Once whole, you have ended your isolation.
One moment is all it takes — one decision.

"Lead us not into temptation . . ." (*Matthew 6:13*) means:
step not into ingratitude.
Because Life is kind.
Life is gracious.
Life is of Love.

The only temptation is ingratitude.

Where there is ingratitude,
there cannot be peace.
Without gratefulness,
we are always trying to release the storm
of our anger, frustration, and resentment.
The symptoms are the same whenever anyone
has not discovered gratefulness.

Watch out for ingratitude.
It is the only sin, the only temptation.
And then you discover how the whole world is caught in it,
everyone, everywhere.

When you are not free of things external,
then you want to own them, to possess them.
And nations build armaments and defenses.
Man's heart becomes hardened,
and you hear the wailing cries of sirens in the cities
and the foreboding news of inflation and recession.

How powerful are these forces of ingratitude!

Do not ever be touched by ingratitude
and so defile your mind which is the Mind of God.
For you and I and God share the One Mind.

Gratefulness

Do not put anything into It which is of ingratitude,
no matter what another has done to you.

If you understood what I have said,
you could no longer have wishes or wantings.
You would be independent.
You would be soaring with gratefulness,
with another sphere of vibrations and energies around you.

Life is good!
Acknowledge the action of Grace.
Be with that thankfulness in your heart.
It is so blessed that no one is alone.

Do not turn your back to Grace to move towards ingratitude.
You have to find that power within you,
to make that decision.
The more we prolong the ingratitude the worse it becomes.
Then the resentment and the casualness
become the breeding ground for hate.
There is urgency only in that.

Are you beginning to see the real meaning of the word urgency?
The meaning of urgency is that
if we do not come to that power of decision,
then we are already with ingratitude.
If we are not with love, then we are with ingratitude.
And you have to make the decision which to be with.

Recognize that the only problem the Son of God has
is that of ingratitude;
that there cannot be "Love ye one another"
where there is ingratitude;
that you cannot speak of love
until you have taken care of ingratitude.

Everyone wants to recognize "holiness," to recognize "God."

You may not be able to recognize God,
but you certainly can recognize your ingratitude.
Recognition is also that you recognize
your own state of being.
If you are not willing to do this,
then you are certainly not interested
in the life of Love,
or in the Love of God.

Whenever ingratitude comes,
see that you are not at peace.
You want things to be different.
Can you then recognize that this is ingratitude?
You are not seeing perfection.
You are not with the Mind of God;
you are with problems.
Problems do not exist by themselves.
Where there is not the "other,"
there is not the problem.

Problems exist only in relationship with a brother.
When you have chosen ingratitude,
you do not have a relationship with God either.
How could you despise, dislike, disapprove
of someone who is an extension of God?

Gratefulness is the solution to all problems.

Brothers or friends are not people you use
as the extensions of your boredom and loneliness.
Friends are those to whom you can give.

Because minds are joined,
your gladness enriches me, as mine touches you.
It is a gift to all mankind upon the earth.
The internal action of gratefulness
enriches something around the planet.

Gratefulness

You will be grateful when you find your own beauty.
And out of that joy is born the new.

If we are moving towards "Love ye one another,"
then we have to conquer, vanquish, dissolve ingratitude,
no matter what the other person does externally.
Gratefulness soars beyond incidents and physical experiences.
When it stands independent of what another does,
then it is eternal; it is vertical.
It is not bound to circumstances.

Gratefulness is independent of one's feelings,
of one's reactions, and logic.
It is not of the realm of the changeables.
Therefore, it is not wrapped up
in the content of hate and suspicion
of collective consciousness.

Where gratefulness is, insecurity is not.
Gratefulness eliminates the very toxins
and impurities in our system.
And you begin to see the false as the false.

When ingratitude and limitation are put aside,
then there is nothing but peace.
A mind at peace has ended ingratitude.

Such a person has already come to action
in liberating himself.
This would be the action of seriousness.
It does not give reality to unreality —
the world of misperceptions.

The means are given to those who are sincere
to overcome the weaknesses.
Life blesses them.
The action is timeless.

Be, therefore, with that peace.

Gratefulness comes to you that you may be so inspired
you want to give love to the brother.
Gratefulness should be one's constant companion.
It is not verbal.

Once you are grateful, you forgive.
Once you are grateful, you no longer judge.
Free of judgment, gratefulness has a clear vision
because it is related to Life,
never to personality.

Where there is gratefulness,
you want nothing from another.

Behold the simplicity
of the poet who said,

> "The heart of a king trembles
> before a man who wants nothing."

What contentment and self-sufficiency therein!
Then whatever you do will be boundless.
It will touch generations to come
for it is the action of eternity.

It is like Thoreau saying,

> *The greatest gift a man can confer*
> *upon his fellow man is to rise*
> *to the height of his own being.*

You sing songs of your joy
for having glimpsed your perfection within;
you are independent of all things external.

Gratefulness

In the disorder of the world
such a man brings an order
that is of Heaven.

Gratefulness, forgiveness, and non-judgment
are very basic.
If you understand any one of them,
the others become easier.

Can the base of our life be gratefulness?
That whenever we deviate from it or get beguiled,
that we can become aware and make that correction?

I am speaking of a state whereby you are related
with whatever you are doing,
whether eating, walking, sitting;
it is a relationship with people,
with work, with all that is.

Gratefulness will not allow us to conform
to greed, to selfishness.
These things would drop away.
Could we then explore its full potential?

Gratefulness thus has the benediction
to free us from our concepts, fears, and weaknesses.

How totally taken over we have been.
"I must look after myself;
I have to get an education if I want a better life."

Everything we do keeps us in the bondage
of our personal opinion.
And within the boundaries of personal opinion,
there is no gratefulness.

The Will of God is with you every moment of the day,

but are you with It?
When you are with It, then you are no longer insecure.
You are grateful.
Then you have the strength to sort things out in your life.

Gratefulness is without preference or choice.
And you are grateful for the Life that meets needs.
It is the poetry of existence
and you soar in gratitude.

Gratefulness!
Then there is the laughter in every cell within you.
Every breath becomes holy.

So, come to simplicity.
Do not look for saints outside of yourself.
You become the saint,
for indeed you are a saint.

What then is gratefulness?
It is a receptivity.
It is being content with whatever Life presents.

Gratefulness is an awareness
in which we see what is ever present
and ever manifesting.
It is a State of Being.
It includes all of humanity.

Gratefulness is the expression
of light within
and your life flowers.
It is when you see the Christ in another.
In Christ, no one is excluded.

To see the Christ in another
is a State of Being in which there are no two.

This action of Life is what keeps everything alive.
And one's heart is grateful
for we are as God created us:
children of joy expressing gratefulness.
And we hear the eternal song
of God's Love for us
spread everywhere.

Part II

Productivity

Productivity

Productivity.
Would you ask yourself,
"What is productivity?"

Would you ever know the reality of it?
Would you identify it with activity?
Are not the mortal versions of productivity
associated with efforts and accomplishments?

Who can earnestly say
that he wants to know what productivity is
because he can no longer live
in the falseness of his assumptions?

Without that incentive, that integrity,
we merely intellectualize
about what productivity is.
Knowing definitions about productivity
is not to know what it is.

The productivity we speak of is a Law.
It is not subject to anything external.
It produces an atmosphere
of the Kingdom of God upon the earth.

We must question all that pretends to be productive.
This passion for truth would silence
all that is of past knowing.
What comes of it is the new.

To question is true meditation.
It liberates you from your concepts,
conditionings, and knowings.
The very questioning itself is productive.

Productivity is
the elimination of self
that stands in the way of Self-knowing.

Productivity is
the plough that digs deep
and uproots all that is unessential.

It is an awareness
that eliminates the illusions of self-deception
and frees one from the system of beliefs —
the manmade world of thought —
and its conformity of unproductive routine.

Productivity is a State of Being
of Divine Origin,
your inner calling,
the appointed role for which all is provided.

Productivity is the Action of Life,
the ecstasy of Creation.

All things in nature are productive
for they are related.
There is not the isolation.
Productivity is
where isolation ends.

The apple tree is productive.
The sun, the rain, the soil, the air support it.
It is related to the universe.

The whole of creation is the joyous song of perfection.
The leaves in their perfection adore the Creator.

Productivity is complete
like the day is complete.
The season completes itself.
There is perfection in it.

Productivity is an expression of eternity.
The wise person relates with eternal Laws.
Thus, the conflict ends and simplicity is.
In that spaciousness
man becomes part of the whole,
a co-creator in perfecting creation
on this timebound planet.
His action affects generations upon generations.

A productive man is one who has peace of mind.
He is a unique being upon this earth
who walks free and pure of all concepts.

Productivity must have
its own Divine definition.
Its action must enlarge our dimension
so that we are no longer subject
to whims and moods.

Could you feel this way as we explore
so that your perception is enlarged forever?

Survival today depends on being productive.
Not the personal ordeal we call productivity,
but the Divine order of productivity.

It is,

Thy will be done in earth

as it is in heaven.
Matthew 6:10

Productivity extends fulfillment.

Fulfillment is an extension of consistency
at all levels of one's being
in which there are no contradictions.
It is an energetic state
that learns in the moment
and thus has the ability to be independent of the known.

Productivity is application of truth.
We cannot separate application from productivity.

As long as we are regulated by the externals,
we are not yet productive.

Productivity is an eternal action
that eliminates external pressures.
In productivity there is no becoming.
It is.

Productivity has no relationship
with one being active or busy.
Have we not always associated
productivity with activity?

Invariably, the root of all activity would be
helplessness, insecurity, and unfulfillment.
They are the intrusions of the preoccupation with survival
and do not relate one to the Eternal Source
behind the appearances.

Activity is not productive.
Productivity is related to eternal Laws
in which there is no insecurity, no dependence,

in which there is not the seeking of results.
It eliminates the future.

There is confusion in our understanding
of activity versus action.

Action is of Life.
Action dissolves the conclusions of personality.
It sees a need, and brings about a change
in man, in humanity,
in an eternal way.

Action does not live by consequences and helplessness.
Action dissolves the sense of inadequacy.
Why take helplessness for granted?

Man owes it to himself to be free and independent,
to be with the action of Life upon the planet.
This is his contribution to the world.

It is man's job never to interfere;
then, Life takes care.
People who are not aware of what Life is
fall into the deception of concepts and time,
and therefore, of insecurity and unfulfillment.

How we live becomes very important
in order to be with a productive life.
We must have a passion to find that purpose.
Having overcome the world,
We are Divinely productive.

And so, we seek to know what it is to be productive,
but what is our knowing or trying to know based on?
Is it not based on thought?

We have to consider what thinking is

in order to come to a clarity
free of self-deception.

We may think we want to be productive,
but does this not create a duality
and a framework of time and distance
in which we must effort to be something we are not?

To find out what real thinking is
would be productive,
for we think in contradictory ways.
It is rather shocking.
We can even justify wars and self-destruction
through so-called reasoning.

Certainly this is not right thinking!

To explore in a profound way,
at the level of reality,
not at the level of opinions and notions,
requires honesty and right thinking.
What ends the words and brings one to silence
is the right use of thinking.

Seeing thought through
to Silence
is the only thinking.
For Silence is free, impeccable and innocent.
It is of another dimension,
independent of thought.

Stillness is wise.
It has mastery over the personal.
In its timeless space,
thought does not intrude.

When your mind is still

you begin to realize that every sound is a song,
an adoration reaching out towards God.
There is something sacred about every sound —
except when it is false.

The false are the ideas of man,
what you "think" you know.

Would you be content to live with empty words then?
It is like sharing paper plates empty of food.

There is a responsibility in talking
about true definitions and not intellectual opinions.
Better to be quiet.
In that stillness you may listen
and find that every sound and every word
has a great significance.

Would you violate the stillness,
believing your beliefs
and thus ignore the Truth?

Ideas assert their knowing ruthlessly.
Hate and fear are born of our knowing.
And someone used six million tons of bombs
on Viet Nam.

There is a lot of so-called "knowing" in that action:
those who manufactured them,
transported them, financed them.
And for nothing but destruction.
Man against man.

Watch out for your knowings.
What you know may not be born out of right thinking.
It may be but conclusions of self-survival
in one form or another.

Several years ago,
someone highjacked an airplane.
Had it not been publicized,
hardly anyone would have known.

Now, every airport in the world
has had to invest in so-called security.
At every gate there are policemen.
Highjacking continues,
backed by the free publicity it receives.

Where is the wisdom
in this accelerated promotion of violence?
Will intellectuality perceive the deterioration of values
placed on human life?

Would you not question
your definition of productivity?
Would you not see that the solutions of so-called thinking
are worse than the problems?
Where is the productivity in this commotion of activity?
Where is right thinking in the world today?

If the premise of man's thinking
is insecurity, unfulfillment, and fear,
what would it produce?

We need to be productive.
Productivity is the awareness
that perceives the false as the false.

Once you see the shortcomings of sightless eyes,
you may awaken to another sight,
another value, another Reality.
This Reality is a gift of God
and it never separates nor divides.

This awareness is never isolated,
and it is out of this wholeness
from which clear thinking comes.

As long as one is insecure and afraid,
ignorant of love,
there can be no such thing as productivity.

We have become self-destructive human beings
and somewhere we must hate ourselves.
For when external affluence comes into being,
we seek more outlets, more ways of escaping
from what we really are.

There are shops by the miles,
a thousand ways of forgetting yourself,
with beer, cigarettes, sex, drugs —
the abuse of bodies.
Anything to get away.

Do you not see that?
Is that not a fact?

Most people are poor in affluence
and no one can afford simplicity.
Can we question the detriment of affluence
to man's integrity?

How little we know of simplicity,
of its space and wisdom
and of what is Divine.

Can you imagine the extent of unessential things
which are produced by this affluent society?
What an indulgence!

One is either an employee

or a businessman.
Can you imagine having as a profession
being an employee,
doing paperwork,
or being a clerk somewhere?

Have you ever looked into the consciousness
of a businessman?
It never departs from profit.
Even when he talks,
you hear the clink of coins in his voice.
Loss and gain.
Where, in this nature of consciousness,
is man's need for truth
beyond intellectual perception?

Intellectually, we know so much.
We can even say we are made in the image of God
without knowing the truth of it.
It is a truth,
but it is not your truth.

Most of our thought would fall apart
if we had the conviction
never to say what is untrue.

Could a man at peace,
a man who thinks Thoughts of God,
get into the madness
of contradictions and oppositions,
of likes and dislikes?
Could he ever belong to the tribal mind,
the clanship of nationalism and its fragmentations?

Could that be called thinking?
Could thinking dare start with a conclusion
and end in separation?

Separation cannot produce harmony.
It can gain status, even reknown,
but it is still self-centered.

Thinking is merely a habit which promotes
selfishness and greed.
It is an illusion.

Thinking, as we know it,
is the lowest form of intelligence.
Why have we not discovered this?

To be productive is to be free
of belief systems and thought forms,
cleansed of self-deceptions,
attachments and petty knowings.

Right thinking would confront all self-deceptions.
It would see that most of our achievements
are a promotion of self-centeredness
and are usually at the cost of someone else.
Do you see the deception of status
inherent in our thought?

We glorify self-centeredness.
But is not self-centeredness always preoccupied,
burning itself in its own energy?

The person who dissipates
this sacred energy through thought
is an irresponsible person.
This same energy can also become
the wisdom that removes the deceptions.

Productivity has no relationship
with irresponsibility.
Responsibility then is when you will not let

anything enter your mind that does not belong there.

So, where would you go to learn to be responsible?
Do the universities teach you?
They will give you a skill to make a living.
The cleverer you are,
the quicker you get the things of the world.
At the "me and mine" level,
productivity is to be questioned,
for there is no love in selfishness.

Have you ever questioned
if productivity can exist without love?
Love is, wherever it is invited in.
Love is the Given.
And the Given is made accessible to each one.
Can you receive it?

> Love is productive.
> Fear is not.

Fear is of the world.
Under the stress of the present day,
one either becomes indifferent or violent.
The violent ones take care
of the indifferent ones.

Not having discovered that we are not of the world
is to die in ignorance.
HE has said,

> *That which is born of the flesh is flesh;*
> *and that which is born of the Spirit is spirit.*
> John 3:6

Only when you have something to give
will you know peace in this world.

If you do not have peace within you,
whatever you do is a distraction.
When you have peace within you,
you learn to care.

In times of crisis,
it is your own inner strength
that will care for another.

When we can care for another,
then we bring something of Heaven into being.
This bringing of Heaven into being
ends animosity, tension,
and the ruthlessness of advantages over another.
Where there is love,
I cannot cheat you.
I care for you.

Love is a State of Being.
It is not for an object, or even for a person.
Love is never divided.
It just is.

Love knows that there is only one Reality
and that Reality is God,
ever productive
and manifesting all that is.

Love is not subject to circumstances.
It does not recognize "good" and "bad."
It is the source and master of all things.
Since it is the source of Life,
how can it hate one and love the other?

The source is beyond the appearance, the effects.
The source is complete unto itself.

Life is the source.
When we are not related with Life,
we are related with the effects.
Relationship is only
in the awareness of the One Life.

The energy of Life that keeps us alive
is very productive, is it not?
It is the involuntary nervous system within us,
and the organs, that keep us alive.
We could not live a moment
without the circulation of the blood.
The eyes that see have something of that miracle.

What a gift to relate
with the Divine Intelligence that sustains.
Have we lost all sense of wonder?

The still mind knows appreciation.
Stillness is there when you are attentive
and not dissipated by thought.

If we really knew the right use of thinking,
we would have time for stillness and silence,
would we not?

Try to put aside one hour in the morning
and one hour in the evening for meditation
and you will witness how activity
intrudes upon stillness.
You will discover how helpless you are.

What wisdom it takes to come to the simplicity
that awakens our Divine faculties!

Stillness is no longer a need
that you can easily meet.

Wishing and wanting have become
the reality of an indulgent mind.

If we really knew how to think,
the very thought would be authentic
and would always return to stillness.
But because we do not have the energy,
we conclude, compromise and stop short.

We usually do not see anything through to the end.
Seeing it through
means to take it beyond thought
and come to stillness.

Stillness activates all the centers of the body.
Stillness opens the temple of the body within you.
And you are no longer the son of man,
but the Child of God.

Whether it is the religious perspective of life,
or it is the way of devotion,
the action starts at the level of the heart.

What is from the heart moves towards perfection.
What we do from the brain always finds expediences.
And this we call "progress."

The action of the heart changes a person.
It would lead us towards,

Love ye one another.

What is of the heart becomes intrinsic, complete.

Being productive also has its own intrinsic merits.
There is a completeness, a satisfaction
in what is intrinsic.

Intrinsic means doing something for its own sake.
But the word intrinsic has an even greater horizon,
greater meaning.
It is an eternal state,
beyond the realm of personality.

When you find what you want to do,
and you do it with your whole heart,
that is intrinsic.
Then all the energies are focused in you.
It is joyous and most creative.

Moses parted the sea.
This is the energy of co-creationship.
Behold the tremendous focus of vitality
in whatever the person is doing!

These energies are the extension of holiness
and relate you with Divine energies.
Extension takes place which is beyond the body,
beyond thought, beyond the conscious mind.
Extension has nothing to do with time.
It is continuous.

You cannot be intrinsic until you are related
to Divine Intelligence.
You are then an instrument of the Mind of God.

He who is intrinsic is satisfied
with the art of living.
He lets all things be exactly as they are.
Intrinsic lives cannot be distracted,
cannot be deceived.

What makes you dependent cannot be intrinsic.
And we are mostly dependent on the physical senses
which we obey most diligently.

Find out what it is your mind obeys.
It either obeys the Mind of God,
or it obeys the life of the body.

The senses produce selfishness, attachment, self-interest.
Competition and ambition are one name.
Yet it can be said that
there is no healthy competition.

What is intrinsic has no seeking in it.
It does not seek because it is already content.

One who has no needs is intrinsic.
But he meets the needs of others.
And there is a crying need in the world!

Would it not then be intrinsic
to awaken to your own potentials?
The awareness of your potentials would know
what productivity is.
Partial awareness cannot.
At the partial level of awareness,
there is fear,
there is insecurity,
and there is unfulfillment.

Nothing seems to bother us unless it is painful.
But is it not painful to be insecure,
to be unfulfilled and constantly looking for amusement?
Perhaps we have gotten used to it
and accept it as a way of life.

We have to recognize the limitation of thought
that promotes "me and mine."
Within the framework of the known,
where the "me and mine" flourishes,
we are busy and active,
but not productive.

Seeing this,
you are of God's perspective
which imparts strength;
manmade rules do not.
And you are grateful for the strength that is given;
there is Grace.
Thus liberated, your life becomes productive.
It is creative
and what you do is of love
for you are now a co-creator.

What is the actuality of co-creationship
beyond the verbal answer?
As a co-creator,
you want nothing from another.
Yet you always have something to give.

But first you have to die
to the false notion of "me and mine"
with its insecurity and unfulfillment.

Without inner stability,
obviously one is not productive.
Inner stability is not externally regulated.
It is not even something one learns.
But it comes from the realization
of what the reality of being co-creator is.

So, what does it mean to be co-creator?
It must be the highest and the most sublime expression —
one with the Essence of God!

When you discover that God, the Creator,
wants your input in creation;
then to die to the "me and mine"
becomes not only effortless,
but a benediction.

It is the best gift you can give
to yourself and to mankind
for you have understood what productivity is.

We are children of God.
He is our Source.
And there is only the one Reality that is alive.

At the personality level,
idiosyncrasies and personality clashes exist.
But we do not have to relate to one another
at the level of physical senses,
the level of false realities and judgments.
This is but the world of reactions.

If we could see through
the separation between "you" and "me,"
that we share the same life,
then whatever you do would not affect me.

Beyond right and wrong
there is a place of love.
You are part of its peace, uncontaminated.
You give of what you have
because you are independent
and cannot be regulated by reactions.

Begin with self-honesty.
Honesty would see that helplessness is something
thought has manufactured,
but it is not a reality of life.
What a discovery that would be!

Helplessness is your own illusion,
and is of your own making.
Beneath the helplessness is hidden pride and boasting.

Free of "me and mine,"
you are boundless.
You are beyond space.
Everything in creation would obey you,
for you become a law unto yourself
when you have seen your own glory, your own holiness.

If you could but realize
that all the planets play their part
in your existence here.
You are important.

You could no longer fit into problems.
You would communicate to another person
that same inner peace,
that same inner clarity,
and dissolve his problems.

What you impart is something different than thought.
You impart love.
You impart that boundlessness.
You bring man to his right perspective.

Once you reach that State,
all you can do is love everyone
because you know
they are as perfect as you are.

Love is creative, energetic.
It is productive to bring another
to the discovery of his own perfection.
This is called relationship.
This is the action of Love.

The only thing that is productive
is to bring the Kingdom of God into your life
and into the lives of others.

You will know the Kingdom of God in you
when you care to bring it to another person.

What a miracle is right relationship!
Everyone is capable of it.
Do not underestimate yourself.

There is urgency in the world.
These are times of crisis.
Behold the consequences of affluence.
We must put away our deceptions,
outgrow circumstances and pressures,
and discover our own reality
so that we can be a light in the world.

The religions of the world have, at times,
gone so far astray,
downgrading the human being —
made in the image of God —
with the authority of a concept
worked in their favor.
Each religion has wanted a God for itself.

Separation and division in any form
cannot bring harmony.
Are you beginning to see
that the human being seldom really thinks,
but merely follows?
And there is no compassion in it.

Compassion does not annihilate another,
for the reality of compassion is that of Oneness.
In compassion there is no fear,
no hate, and no separation.
It knows not the belief systems of right and wrong
which are of the earth.
"Love ye one another"
is the ultimate Law.

Just one commandment.
Do you not see how simple it is?
Once you have found the peace within,
you will never pursue success.
You have no use for ambition.

Otherwise, thinking remains
geared towards things of the world.

The man of God owns nothing.
He seeks no success.
External achievements have lost their meaning.
They are for people of the physical senses.
He does not value what is valueless.
And that is productive.

The man of God is the king of the universe.
"How does he make a living?" one might ask.
Extremely well.
Awakened to his Divine faculties,
he is surrounded with the vitality of Heaven.

He will take care of the body needs.
He will eat the right food.
And whatever work he does
will have an integrity to it.
He has the time and the space.
He has the resources of Heaven
to meet every need of humanity.

There will never be a time
when such a person
does not have something to give to another.

The issue is that we must evolve
something within ourselves to give to the world.
Thus, the source of it is something
other than a skill.

The instant we are the extension of the Origin,
we are liberated from the fear
and the assumptions of the senses.

These are our potentials.

Productivity does not lose sight of the fact
that harmony is more precious than objections,
quarrels, points of view,
and the evaluation of right and wrong.

Similarly, gratefulness is
an expression of the joy in you
for what has been bestowed.
A living gratefulness so rich
that no one can take it away.
No matter what another does,
you could drown it in your joy,
dissolve it and transmute it.

Stillness can cope
with all and everything.
Sorrow could not touch you,
and nothing could contaminate you from the outside.

The joy and happiness
that are not insecure,
nor dependent on the externals,
are productive.

Productivity is,
when there is no unfulfillment.

> *Thy will be done in earth*
> *as it is in heaven.*
> Matthew 6:10

Thus, the forces of Heaven and of the earth
combine in you and become One.

We can become an extension of God
rather than being externally regulated
and problem-ridden.
This is our inheritance.

This productivity is not personal,
but an extension of
"Love ye one another."

It's inexhaustible energy responds
to someone else's child, or leaking roof,
and meets the need.
Being truly productive is being part
of this inexhaustible energy.

For the most part,
man knows the energy of conflict
between people and nations
seeking peace through war.
The energy of friction has its consequences.
A seed must inevitably reproduce what it is.

So much of what goes under the name of productivity
is the efficiency of fear and greed.
We may call this progress.
There is no end to interpretations
which remain ignorant of truth and compassion.
Knowing all, yet knowing nothing.

Does not our education prepare us
to fit into the social structure of this world?
And the social structure believes in war and violence.

What place has innocence

and its vulnerability in our times?
The poet calls it
the syndrome of educated ignorance.

The voice of a man who is liberated,
is authentic.
His words, being realized,
have a life behind them
and are ever productive.

> Authentic words have an effect
> beyond intellectuality.

They carry a blessing with them.
And they will be heard.

Christ's response to the fragmented world is,

> *Father, forgive them;*
> *for they know not what they do.*
> Luke 23:34

Love sees only God.
It is love that is all-inclusive.
In Life, no one is excluded,
even though they be of different religions
or thought patterns.
The different is not wrong.

This is applicable to all ages
and to all races
and is productive throughout time.

The man of God comes to the earth,
makes peace with time
and affects it with his eternity.

When you are true to your productivity,
you will have no thought of tomorrow,
because you will not deviate
into thought concepts.

Deviation into cares of tomorrow
is a violation of the law of productivity.
All is provided.

Productivity is the light of Heaven
that tranforms one's life.

Prepare yourself to undo the conditioning.
Become attentive.
Shed the judgments.
Free yourself from reactions.

Simplicity and relaxation are the keys.
You will see wisdom expand its horizon
and change the quality of your lifestyle.

The energetic state of stillness is content,
for it has found trust and faith.

Trust is what gives one the strength.

Without trust,
there can be no honesty.
That is a Law.

Honesty is not possible
without the stability of internal stillness.
Neither is trust possible
without harmony with your environment,
with your self,
and with others.

Faith has nothing to do
with dependence or belief.
Belief is reliance on another.

Faith is independent of the past.
It is as alive as the Present.
Faith is what diminishes any thought
that comes to interfere with the Present.

Many people dissipate their energy
in destroying other people's vulnerability
and having no real relationships.

To serve is something so totally different.
It is to reverse the process
so that everyone is safe around us,
for we have something to give the world.

When we do not take advantage of another,
then we truly are taking a very vital step
in our development.

To each one these intimations are given.
We may never catch the intimation,
or we may even forget it, if heard.

What is an intimation?
It is not a guide "above" giving you instructions.
Intimation is not external.
It is of your own wholeness
that perceives and awakens
the Divine faculties within.

These intimations of the still, small voice
are delicate as dew drops of freshness
that either evaporate unnoticed
or sprout a newness in one's life

that can affect the entire race upon the planet.

The freshness of first intimation —
the language of Silence.
It is not of words, but a purity,
the impeccable space
within the consciousness of humanity.

You cannot isolate it.
It is not yours alone.
It is of the One.

But you can ignore its call
in the sleep of your delirium.

The intimation is the vision.
Heed it, or be absent,
and deprive the birth of a New Age.

The intimation of the still, small voice
dies if forgotten,
or if its love,
which brings Heaven to earth,
is not shared.

Some time ago,
I received an intimation of productivity.
I will paraphrase it.

All through your life
you keep evolving.
And towards the last steps, so to speak,
you cannot go any further,
yet there is further to go.

At that point,
the Life Force intimates

why you cannot go further.
It is more a glimpse of a Law,
though not in words.

You perceive that it is the alive initiative in you
that has prevented distractions to date.
The energy of dedication has brought you
to this point of clarity,
but now you cannot go further, alone.

And now, you must share.

For in sharing is the extension beyond the personality.
Only by teaching will you be given to learn
what you, too, lack.

What a satisfaction to be both
the student and the teacher within oneself!

The student is one who wants the unknown,
the new that ends his conflict.
The clouds of confusion and deception are lifted away
and he is in a different state of being.

In teaching, the student learns what is miraculous.
This means that where two or more are gathered in His Name,
His Presence is there.
And who are the two?
Yourself and the Lord.

Then you discover
that you were neither the teacher nor the student.
In this Reality, there is no teaching.
Truth and Oneness just are.
It is dissolving the learning and teaching.
It is the One Action.

The real meaning of learning
is to recognize who you really are.
Once you have recognized it,
then it is a contribution to all mankind.
In this recognition you have found your function.
You feel so blessed and filled by it
that your heart yearns to help the brother
to recognize his purpose too.
You realize the function of man on earth
is to bring to it the vibrations of Heaven.

The earth provides the body.
Heaven provides the Spirit.
And to the manifested level
descends the Spirit, through man.

The earth feeds and nourishes you
while man brings the blessings upon the earth.
The earth needs your blessing.
This is why Nature supports you, feeds you.
You have a purpose.

Everything in creation
exists to bring man to perfection
so that man may bless creation.

The realization of this brings one
to a new action of productivity,
gratefulness, and forgiveness.

Man brings peace to the earth.
In harmony with the whole,
he walks and blesses nature,
bearing the Heaven he is seeking to find.

The Angels of the land,
of the seasons, of water and of air,
walk before him.

Then we also see that one cannot make it alone,
without the brother,
for isolation is not related to Life.

"Love ye one another"
is a statement of a fact.
Inherent in its discovery is the vitality
and the exuberance to implement it.
It is your strength.

When the application of Love takes place in man,
the consciousness of this planet is changed.
You could never realize love without the action of Grace
and the compassion and perfection of creation.
Grace is ever with you.

In the end, every single person becomes a teacher,
and by teaching, learns.
Giving and receiving are one miracle.

And instantly it is revealed,

> *Ye have not chosen me,*
> *but I have chosen you,*
> *and ordained you,*
> *that ye should go and bring forth fruit,*
> *and that your fruit should remain:*
> *that whatsoever ye shall ask of the Father in my name,*
> *he may give it you.*
>
> John 15:16

What "fruit" would you pray for?

Our fruit could not be personal.
But the fruit we would ask for
would be:

Love,
to give.

In it, there is no "I"
and we can truthfully say,

I can of mine own self do nothing . . .
John 5:30

Thus, there is the productivity
of ending the separation
and the nothingness of unreality.

Once
I perceived being given an eternal bowl —
all pervasive and all encompassing.

As the external illusions of brain activity
began to dispell,
I was told,

"Keep the bowl empty.

Do not put your opinions
and wishes and wantings into it.
Nor the conclusions and assumptions
of another.

Keep the bowl empty."

The empty bowl is the Mind of God
in which all is united.

Now *you* are given the bowl.

KEEP IT EMPTY.

Part III

Time and the Timeless

Time and the Timeless

The relationship between the known and the unknown.

The known is of the brain.

The Unknown
is the glory of the holy Mind of God.

Some things are difficult to explain,
especially if they have never been expressed before
and relate to a profound experience
which is taking place within you.

It is something fresh,
not yet formulated in one's own mind,
something new and real.

What is new and alive makes demands
on both the speaker and the listener.
Within the words
there is the action of silence
which brings one to the State
rather than the known, the preachable.

Can we be so attentive
that the very attention brings our minds to stillness,
where even the pause between the words
evolves an atmosphere of heightened sensitivity?

The mind can only be attentive

when it is innocent of the known.
Innocence heightens awareness.

There is no other way to bring the mind to stillness.
You can dull the mind, get to a sub-level,
and for a while be pacified.
People drink alcohol and take tranquilizers to that end,
and thus, remain intentionally dull.

At that level,
you are trained to fit into a function
but remain self-centered and timebound.
The primary fallacy of such a learning
is the illusion of self-improvement.

Compelled by unfulfillment,
you use your ambition and get caught in ideals and goals.
Then you go to learn from teachers,
from colleges, and from gurus.

It is difficult for people caught in time —
and most of us are —
not to be stimulated.
Fear and insecurity are in the blood of man
and mental pressures are their outcome.
Dependence is exploitable.

While timebound,
we remain caught in illusions
and ignorant of infinity.
Time with its pressures, projects, and ideas,
distorts the eternity of our Reality.

It takes a rare individual to step out
of the conceptual fantasies of it all.
And for that he needs to change his lifestyle.

This is the challenge of challenges.
We settle for adjustments,
but nothing less than total change is required.

Have you met anyone
who is not caught in this dilemma?
Would the one who is in the timeless state,
having discovered his holiness and perfection,
bother about a preoccupation called
learning or self-improvement?

It is obvious that while we are on this timebound planet
and have the brain and physical senses,
we are subject to the laws of time.

Time is the law of this planet.
Things grow; things decay.
Whatever is, is always in that motion.
It could be a rock.
It is still in that motion.

Every single thing is related to something else
for its survival,
for its very being.

A tree could not live without the soil,
and the soil might not have been there without the trees.
Then there is the rain, the sky, the clouds, the sun.

There is a difference between
relationship and dependence.

Relationship is becoming a rare thing.
At the personality level,
there can be no relationship.
We know but the illusion of living
and not the extension of life,

the holiness of eternity.

How few people know what relationship is!
We know dependence for the most part:
dependence on success,
dependence on another for gratification.
And life becomes very abstract.

It is when we put our reliance on the externals
that we are dependent,
bound to inadequacies.

As the externalization increases,
man deviates further from simplicity
into stress and pressured existence.
Violence and tension become inevitable.

The more pressured we are by time,
the more violent and self-destructive
become our tendencies.
Thus, the peace and space within are violated.

The consequences of success and artificiality
are evident in a world where man is isolated and timebound.
We live in an age when virtue and ethics
are forgotten and lost.
And now we have our insecurities
and the fear of consequences.
To struggle to survive within the framework of separation
is to be subject to the deteriorating forces of time.

> Rightness is the law
> before which everything external bows.

The wise discovers for himself what is harmful,
then comes to clear recognition that time itself is violence.

But,
awareness is a place in you
that is neither of the earth,
nor of time.

We have to first recognize that we are subject
to earth forces and human conditioning.
The challenge is to ward off
these adopted tendencies.

Replace knowledge by wisdom,
and your ignorance and worry will vanish.
Do you perceive the difference
in following habit
or following conscience? . . .
The masses follow their lust and desires,
having manifold wants and aims . . .
The masses are busy and chase after progress . . .
But the awakened is unwanting . . .

Lao Tzu

The one who is aware is the awakened one,
united with God.
He is a being liberated from time.

How essential is the fresh dawn
of the Timeless within one
that ends the separation.

In the ancient times,
you went to study from a great teacher
who taught you the miracle of life,
the glory of its existence,
and the Divine wisdom of it.

The teacher inspired you out of the time level
into the Timeless,

and introduced you to your real nature,
which is timeless.

The real teacher is the one
who opens the doors to the truth of your being.

This state is innocent of the ways of the world,
untouched by fear,
inspired by the perfection of each moment,
totally free of insecurity,
never confused.

When you are in the timeless state,
there is no duality,
no conflict.
In this materialistic and mechanical culture,
it is a statement of the eternal on the timebound.

There is not the "other" in the timeless state.
There are no two in it.
There are no names for it.
There are no means to it.
It just is.

The activity of the brain
learns names of things,
accumulating information and knowledge
to take care of the physical self.

But in the Timeless,
you have soared beyond the self
and even the activity of the brain
becomes still and is blessed.

Thought finds fulfillment in silence.

In this awareness,

you see that everything is an expression of Love.
There exists nothing but the wholeness of Love.

Love expresses the sacredness of Life.
It is an internal discovery
that few men have made.

It is not the wanting of love externally.
For it already has and is.
The Being who knows Love
shares what Reality is
and is a compassionate friend of mankind.
He is aware, but not reactive
to the inadequacies of another.
For his is the Path of Virtue,
not limited to personality.

Do you not see the need of this timeless atmosphere
to deal with the deepest issues of man?

In the state of Love,
nothing external can affect you.
Now you affect the external.
You find all things of the earth that are timebound
need your blessing,
your presence,
your grace.
They depend on you.

You manifest Timelessness upon the earth.
You bring to the earth
the glory of Love that you are.

Not to know this is to live in ignorance.
We have paid very little heed to the discovery
that all we learn externally
is but educated ignorance in the absence of Timelessness.

All the scriptures of the world have said
that you are made in the Image of God.
Have you ever questioned your being timebound,
being hostage to your own deceptions?
That this cannot be it;
there must be more?

The known is of the brain.
Could you question the limitations of the known
and the fragmentation that it causes?
Could you then come to a state of being
in which you will never compromise?
Compromise can never know the timeless state.

In the absence of trust, however,
there are only compromises,
as absence of love has its consequences.

In trust, compromises end.
In love, consequences end.

When the truth becomes clear,
you discover there is no other way
than that of Rightness.
Rightness is whole and non-compromising.

If you take life casually,
then you get by with making a living
and solving problems that are self-created,
becoming a businessman or someone's employee.

Is this not third-rate status for a Child of God,
to be tied to the schedule of a boss —
a boundless being in the prison of time?

Depression and loneliness originate
in the separation from God —

such loneliness, such depression
that we constantly need something to gratify ourselves —
a thousand indulgences in a culture
that predominates in restlessness and pleasure,
providing means of escaping the self.

Humanity today spends more time and money
on outlets of gratification
than on any other single factor.
With a myriad of distractions,
we come to die never having lived.
For all that is of time must die.

Behold the culture of shattered silence
and its children!

In this world,
controlled and crippled by loneliness,
exhausted by the daily routine
and the demand for distractions,
man becomes helpless and turns to violence for solutions.

Now the urgency is being felt
for the vibration of survival is upon the age.
We must correct the lifestyle
of dependence and helplessness.

The time has come when the individual must question.
It has been said that learning
can only be recognized by its results.
Not to have discovered that you are of God
is to have learned nothing.

Either you recognize you are of God
or thought identifies with self.
This causes confusion and distortion
for the self may not want to question itself.

You have to be discontent enough
to go beyond the self and silence the thought.
Here the interpretation ends.

We are children of belief systems.
Who has a thought of his own?

We have a belief that it takes time to learn.
But awakening is instantaneous
and independent of effort.

Can we end the traditions of self-deception
and demand of ourselves direct discovery?
The most difficult challenge
is the unwillingness to see one's own illusions,
the unrealities of one's own life.

How few ever rise to conviction and integrity
and stay with rightness.
How many fall for success
and become the destructive residents of this earth.

It requires a reversal of thinking
to be the extension of the Grace of God
rather than the fear and insecurity of the earth.

To outgrow the unreality of experience
that makes one captive of physicality
is to outgrow the unreality of time.

The Law is:
all we have to do — and must do —
is to bring the old to an end.
This is our responsibility.
The New already is.

The appointed function of each human being

is to bring the Kingdom of God to earth
and to live by,

Love ye one another.

"Love ye one another" is natural and effortless,
like the fragrance of a flower.
How could there be any struggle or difficulty
if it is effortless and natural?

Your very essence and source is Love.
You do not have to tell the tree how to grow
or the fruit how to ripen.

Hopefully it is beginning to dawn
that the action is of Grace,
and the Lord does His work.

The wisdom and the grace of "Love ye one another"
is the most glorious, the most essential
of all commandments.
It is all we need to learn,
if learn we must.

What place has your external learning
when you have been made aware of "Love ye one another"?

But who ever listens?
Who ever hears?

You have heard a truth
only when you live it.

Have you ever wondered why He would give
but one commandment which has
no struggle, no goal in it?
Can we step into the timeless state

to see the truth of it?
For it might not be an ideal.
It may well be a fact.

"Love ye one another" is your natural state.
See that if you had to achieve it,
you would attach it to ambition
and murder another in the name of religion —
which is what we have done throughout history.

The timeless state first sees the self-deception:
"I believed it was difficult to love another;
I did not even love myself;
I love gratification;
I love to be at the point of advantage."

Love knows no advantage.
It is not a barter.
It is a law unto itself.
Once you know it,
you are never affected by anything,
yet you affect all that is.

A man crucified could still say,

> *Father, forgive them;*
> *for they know not what they do.*
> Luke 23:34

Behold the statement of Love!

Nothing external can affect Love.
If you can justify not loving another,
then you do not yet know what Love is,
nor your real nature.

One begins to see the importance of "Know thyself."

But we get busy improving and educating the self —
the commotion of brain activity
which produces yesterdays and tomorrows.

It is thought that manufactures time.
Time is the projection of thought.
One instant of insight can take us beyond thought.

When the mind is still,
there is only the Present.
And the Present is eternal.

What we know does not apply.
What we know could prevent us
from being in this moment.
What we knew half an hour ago
is an untruth and a lie
to the living moment of NOW.

If your brain is still active and verbalizing the known,
then you are far from the freshness of the moment
that demands the urgency of stillness.

The yesterdays and the tomorrows of thought
are the illusions in which we are caught.
Every moment is perfect,
for every moment is timeless.

This is the receptivity that silences the mind
and beholds the shining face of God.
For He alone is.

There is a relationship between
the timelessness of the mind
and the timebound brain.

The physical part of our being

is made of the earth,
nourished by it, lives by it.
It remains subject to time.
For time is real to the physical.

The earth needs man to bless it
with the holy vibration of Timelessness.
This is man's function.

To know the relationship between time and the Timeless
requires wisdom,
rather than knowledge.
Wisdom and learning are quite different.
Learning is always of things external.
We need to learn at the realm of sense experience
for our physical welfare —
whether berries are poisonous or edible . . .

But wisdom has nothing to do with the external.
Wisdom is another sanity.
Without wisdom learning becomes insane
and manufactures nuclear submarines.
It is a fragmented mind that produces hate
between race and people.
The brain is obsessed with the memory of blame
and prejudices towards other people.

Behold the state of the world —
the Mind of the Age.

Would it comprehend the purity of Heaven?
What would bring it to serenity?

Deprived of relaxation and the rhythm of the day,
the impurity of the blood circulates
through the brain and produces anxieties
and tensions in the nervous system.

Health is basic
for the human body is the altar of God on earth.
Could you question the medical definition of health,
and discover it from God's perspective?

Food is a factor.
Each person must select the food that is right for him,
for diet is determined by one's state of being,
evolvement, and direction.

If one is with his appointed purpose,
there is no conflict within.
What would the diet be of a person free of conflict?
Obviously, the quality of his diet would be different.

Let us be concerned with the change of values
and right livelihood, and not just the diet.

Discipline is a factor.
What would it be to know the miracle
of sleep and relaxation and simplicity?
Simplicity is the key to a new way of life.

Heed the wisdom of the Chinese Proverb,

> *We never buy more than we need.*
> *We never need more than we use.*
> *We never use more than it takes to get by*
> *till we learn to need less.*

We live in a culture that abides by the law of wishes,
where the majority of the world population
produces unessential things.

What most people relate to and are limited by
are their desires and wants.
In the absence of simplicity,

the unessentials assume enormous importance.
The human being is victimized
and left with a life of consequences.
Yet we are responsible for the purity of peace within.

Humanity has gotten out of gear.
It not only has become insensitive,
but self-destructive through fatigue.
Even the planet is worn-out.
The fatigued world resorts to armament
and there are flags for which you die.

Nations now spend over six billion dollars
on armament annually.
There is a profound lack of wisdom in the world
that believes in fear and wars.
The vibration of survival is upon us.
Can we end this momentum —
the destructiveness of external unrealities?

The man who realizes the fallacy of cause and effect
does not get involved in organized causes.
He sees that the knowledge of the external,
without wisdom, is incomplete.

Wisdom introduces one
to the timelessness of one's self.

Therefore, nothing of time
need ever own you or frighten you.
You are made in the Image of God,
and God is not timebound.

We are the children of the Almighty
and our true nature is that of Love.
To be with Love and Timelessness is never to be
in the bondage of circumstances.

You and I must come to an inner awakening
to discover the wholeness of our Reality.
It is difficult for the human being to let go
of what he thinks he knows.
Preoccupation with what we know
is given the highest value.

But where is silence?
Where is serenity?

"Disquietude is always vanity,"
exclaimed St. John of the Cross.

Spare time for aloneness,
some moments of innocence,
some moments of non-intrusion of timeboundness:
these will transform you.

Meditation is an awareness
to which all things are external.
Surely we have all tried to sit quietly at times
and most likely have convinced others
of how nice it is.

There are courses on meditation
which promote some activity or another.
They tell you "how" to do it.
It is absurd that man would have to learn
to be still and silent.

See how deception takes place
and does not allow one moment of peace.

There is only one peace.
The Peace of God.

When a man is at peace

he perceives the holiness of God
that surrounds every object.

So in the evening, let the serenity of twilight
end your duality and make your life a prayer.
Gather yourself in the solitary hour of wholeness
for the vertical aloneness
that steps out of activity
and soars to heavenly heights.

Prepare for the miracle of sleep
where physical consciousness does not intrude,
but the temple within you
opens its gate for you to enter the peace therein.

You are not of the earth.
Your reality is Divine.
In sleep you awaken unto the realm of God.

Let your awareness spread its joy
into the soft light of dawn —
upon the world of appearance — and give it peace.
By you, His Will is done on earth as it is in Heaven
and the gladness of your heart sees the glory of God.

Everything is perfect in its own reality:
the flowers, the leaf, the night, the stars,
the function of the knee, the palm of the hand.
Everything is complete.
A child is not an unfinished man.
He is ever complete.

When you see the perfection in all things
you are inspired.
Your mind comes to peace
and your heart is at ease.

Inspiration brings new space:
a freshness of silence
where time and the Timeless meet.

You have heard enough.
No one can teach you further.
You can directly experience it,
if that is what you want to do.

Be touched by your own Timelessness
for it is not something another can give you.

God has already given it to you.
It is not something you achieve.
It is a discovery that it is always there,
but you have been absent.

You could give yourself no greater gift
than a moment of your Reality.
Fall in love with the God within
for that is what you are.

To come to the Present,
to come to the Timeless,
you need but become aware of what prevents you.

Life is sacred.
In its holy perception you are made whole.

Could you then bring upon yourself a transformation?

And if you do not,
who will?

Part IV

Silence

THE SILENT WAY

Choose once again. For it is given you
To trail the peace of God across the world
Without exception. Every child receives
The gifts you bring, and men and women turn
To you in thankfulness. With joy are you
Accepted everywhere. For you have come
Only to bring Infinity's appeal
To those who are as infinite as He.
You come with memory of God in you,
To waken this same memory of God in those
In whom it seems to sleep. The world would die
Without its saviors. Do not, then, deny
Your proper place. For Christ has called to you
To follow Him, and choose the silent way
That brings you to eternity today.

from *The Gifts of God*
by Helen Schucman

Silence

Silence is a State of Being.
Its first relationship is with God.
Everything else is secondary.
Silence is a statement
that you are not of the world, not of the flesh;
that you are purified from wishings and wantings.
Spirit knows no wants.

What, then, is the relationship of Silence with creation?
It is the movement of a prayer and a state and a stillness
with which man blesses all that is created
and all that is part of his life.
He becomes whole.

The silent state is one in which one owns everything
and therefore has no wishes for anything.
Behold its dignity.
Owning everything, it wants nothing.

One can hardly describe in words what Silence is.
But let us see what the action of Silence is.

Silence is an active undoing of wishes, knowings,
conclusions, urges, habits, and conditioning.
Thus, Silence purifies itself.
It is an involuntary cleansing process.
You give it the space, and do not interfere.
Silence, when not imposed, has its own energy.

Silence is a statement of ever-renewing gratefulness.

It is a peace as big as the universe.
Nothing is outside of it.
How can you be alone when you are related with Life Itself?

Silence is afire.
It is vigilant.
Its one day is never like another.
The energy of Silence brings one
to the stillness of the Absolute.

Silence is not self-centered.
It is a sensitivity that is not regulated by thought
or the constant craving of the body senses for experience.

To want nothing is consistent with Silence.
To wish is to violate contentment.
In aloneness one misses nothing.
It is complete unto itself.

Not being touched by loneliness
is the proof the Silence is genuine.
Silence and aloneness have an affinity
for they are both aspects of the One Wholeness
in which all is complete, holy and glorious.

Silence is ever New.
It is the discovery of insight at work.
Its energetic Awareness undoes personality
and relates one with all that is.

You learn from what is and respond adequately.
It is very different from projecting a knowing
and then implementing it.

Within Awareness there is an observation
which is never reactive, but wise.
It sees the falseness of thought behavior,

beliefs and patterns,
but is never unaware of the godliness
of the human being.

When one is related with Life,
there is only the appreciation of Creation.
All is provided if you are emptied
of your own knowings of unfulfillment.

Silence is a Law.
It knows no limitation.
It is not subject to the changeables.
It is not external silence.
It is the Silent Mind.

Ever effortless, ever energetic, ever present,
attentive and joyous,
the action of Silence is direct.
Thought becomes sparse as pressure ceases.

What you are begins to reveal itself.
The question of what to do with your life —
of helping, planning, needing people — disappears.
These are but the deceptions of helplessness.

One might even ask,
what is the relationship
between Silence and seeking to be silent?

Silence is to realize: I am that I am.
To know Silence is to know the art of living.
Seeking is an activity of "becoming."
It has no relationship with the silent mind.
The activity of seeking Silence is a deception.

To be silent at all levels of one's being
requires honesty, not ambition.

For Silence has nothing to do with achievement.

Could one consider if it is possible
to transcend so-called "learning,"
the pursuit of "becoming,"
and be still?

Many basic questions would arise within.

Is it possible for man to deal with the power of insecurity?
Does one dare challenge all one knows that is not of Truth?
Can fear, which dominates man's life, be silenced?
When did fragmentation begin . . . where . . . how?
How does separation maintain itself?
Would aloneness end dependence on all that is external?

Yet man is so afraid of being vulnerable.
He remains regulated by resistance, sensation, and opinion.
When one discovers these forces
making for wrong-mindedness within oneself,
one wants to confront it all.

"Know thyself" soars in its importance.
Do we have the space to give
to allow Self to reveal Itself?
Could we put all opinions and knowings,
even wantings, aside to learn what Silence is?
For if the brain interferes,
we will never know Its reality.

Not to interfere means to have discipline.
Discipline is not imposed.
Discipline actually means to be attentive, not controlled.
What man knows for the most part is but conformity
in one form or another.

The unpredictable, the new,

is as yet unfamiliar to us.
What we know is the intrusion of desire,
of attainment and of gain.
This, obviously, is not Silence.

The worst of all indulgences, however,
is the desire for more of the same.
Memory and desire co-exist and are synonymous.
Gone is freedom in the supremacy of thought.

Silence is a space far more important
than thought, than activity.
Activity is based on time.
Silence is independent of time.
Activity has its own pressures.
Silence has none.

And we live in times when we are under the stress
and pressure of circumstances and of time.
We have all but invaded Silence out of our existence.
Tremendous artificiality has come into our lives
with the absence of Silence in our daily living.

What is this lack of fulfillment in the world?
Behold what chaos it has brought.

All through the ages, man has tried to come to terms
with his two selves:
the personal self and the God-created Self.
This is the basic issue.
All the scriptures have pointed in that direction,
but who has dealt with it?
Silence is the premise and the foundation of "Know thyself."

With Self-learning, sanity dawns too.
And what a joy is inner correction!
You become your own teacher.

The awakening of potentials begins
and dependence loosens its authority.

How few have made the decision that their first love
is for that which is eternal.
Therefore, it is not of the body.
This decision would cope with the urges and impulses.

Do you feel that you are His Son?
Do you walk that way?
Then nothing of the earth could entice you
because you are out of it all.

How vast is the truth of what we are.
How sacred is each one of us.
How holy are we.

God's Son is apart from time,
united with His Father forever.
It is his love for the Father
that enables him to overcome the world.

If you could but see the reality or holiness of one person,
it would introduce you to all humanity, to all life.
There is only the One.

The One Son is <u>you</u>.
And it is your responsibility to realize this truth.

Silence sees that real memory
is the memory of being non-separated,
where wholeness touches what is eternal,
beyond the personality.
Silence is independent of the body.
Personality is at its command
and implements vision on the earth plane:

Thy will be done in earth
as it is in heaven.
Matthew 6:10

There is no peace at the external and horizontal
level of existence.
To know the Peace of God,
we have to live consistent with Eternal Laws —
His will on earth.

The one who is factual is a light unto himself.
He is extremely wise.
The wise is never deceived.
To others he introduces the illusion of thought
and the reality of fact.

Such a man is most needed.
This is his function, his destiny.
Such a man is you.

The fact is,
you are made in the Image of God.

Do not underestimate yourself.
You have the resources.
For you are a Child of God,
of Love, of Light.
You are an extension of perfection.
Do not assume that you are imperfect,
and then seek perfection.

We have to open other doors
in which there are no wantings, no unfulfillments.
We must discover the wisdom Stillness has to offer
to make right use of the body
and right use of the brain.

What wisdom would it take to bring the brain to harmony,
to cooperation, to lovingness?

The true nature of the brain is to be still
and to bring things to order.
The brain is quite intelligent.
It asks questions in order to bring things to completion
and then to be still.

When the mind is still, the body is happiest.
You can sit for hours without moving.
The body, being content, cooperates with that Stillness.

A still mind means
the brain has ceased and the light has dawned
and nothing is outside of it.
Everything is within, and you are vast, unlimited.

The brain that never thinks of anything other
than what needs to be completed
is a friend of the world, of itself, and of rightness;
in no time, it is silent.

Can you see how the brain has the capacity
to be renewed and to be cleansed?
What cleanses it — what renews it — is Silence.

We have a responsibility to come to the Stillness
that renews man.
For the brain deteriorates when it is constantly
worried and anxious.

It is your responsibility to live a life
in which the body can relax and come
to its own stillness very quickly.
Have you paid attention to this?

Can we come to a space where the brain does not intrude?
You would have to be relaxed and attentive.
The Living Moment of the Present
cannot be touched by the brain.
The brain is of yesterday; it is planning tomorrow.
Brain activity is physical.
All of its knowing is of limitation.

The Living Moment is timeless, untouched and immaculate.
It introduces you to your wholeness,
and nothing changes it.
All of man's learning has merely been a barrier to it.
The Living Moment cannot be related to this planet alone.
It must be related to the whole universe.

Sitting quietly makes us receptive to the Living Moment.
Then you discover that there is only God,
and nothing else exists.
Who is there to fear then?
You can only love.
To love God is never to be absent from the Living Moment.

We have to bring our thought system to Silence.
This movement of coming to Silence
is even swifter than judgment.
It takes another kind of energy.
We have to live differently.

I am talking about a state in which
thought is so intensely alive
that you are beyond personality.

You are eternal; you are boundless.
You are a light unto yourself
that thoughts have not blocked.
Then you are free of all conclusions — innocent and alive.
That alone is religious because it is not a dogma,

not a belief.
It is a state of Being.

> Nothing is as powerful as Silence.
> It is the height of sensitivity
> and the greatest responsibility.

We have tremendous potential that we cannot discover
until the brain is silenced.
This is the beginning of wisdom.

We cannot ignore our potential.
If we do, then we become dependent.
This is where helplessness is born.
When a man ignores his own potential,
he becomes victim of his helplessness
and encourages exploitation in the world.

When the brain is still, the Light of Heaven
comes to burn all that is false
and awaken you to your own purity.

> Silence is a fire that burns all impurities
> that doubt activates.

Your stillness becomes part of the One Mind
which is of God.
You discover that this State has always been with you,
but you have been absent and preoccupied.

The One Mind is that in which
there is no separation between you and God;
between brother and brother;
between truth and truth.
It becomes one.

Where there is an opinion,

separation has already taken place.
Opinion can never know oneness.
Opinion must isolate one from the other.

It is the action of Love that ends separation.
Love is not a word, but a State of Being.
In order to know what Love is,
we have to come to the eternal Self that we are.
It encompasses all the wisdom of Heaven,
and all the virtue.
It is the truth of,

> *I and my Father are one.*
> John 10:30

This is a statement of Silence.
In those words is the vibration
to bring you to that State.

To make space for this glorious and wondrous Silence,
is the first step.
To relate with what is eternal,
the body first comes to being comfortable,
the brain comes to completion,
and Silence begins.

Within the Silence there is another action.
All the knowings disappear.
It discovers a freedom from personal bondage.
Then you can say you are free.

When you are free,
you do not need words to confirm it.
You have touched upon the Unknown
and seen the triviality of the known.

Your purity blossoms

and you carry it with you no matter where you go.
It does not have a religious name.
It is a laughter, a joy within you forever.
Whatever you do, that gladness will be there
and it will be meaningful.
You have something to give to the world.

In the Silence, you create your own atmosphere
and you go to the world with your peace.
Then you affect it.
This is transformation.

You come to an intensity of Silence that is your own.
No words touch it.
No interpretations touch it.
It is a state that knows no problems.
Out of that Stillness the new is born in your life.
What is new is of God and is found within.

There is a Voice to be heard
when all fragmentation has ended.
It speaks of what is eternal,
something that has never been said before.
It is always new.

If you have the ears to hear,
you receive the newness, the holiness,
and the freshness of it.

Only the internal can know what is eternal.
Words will never know it.
Thought cannot go to it because thought always wants.
Thought has relationship only with the external.
Where there is no wanting, there is no thought.

Question your own thought,
knowing that Silence is superior.

Confront thought with innocence, free of knowings.
Innocence is whole; thought is partial.
The whole can never be defeated by the partial.
This is a Law.

When thought ceases,
you will find more than thought could ever want
already within you.
It is the discovery of wisdom, so intensely alive
that its light covers the world.
You are unlimited.
You are as pure as innocence itself.
You are the breath of Heaven upon the earth.

You are body, mind, and spirit,
and must give priority to things of the Spirit.
You have to be still
so that you can receive something of Heaven.
There are no means to Silence.
It is.

How energetic is Silence!
Probably the most energetic action man is ever to know.
Gratefulness is an expression of that Silence.
It expands towards all that lives and breathes
and shares creation with you.

It is a state that is absolutely impeccable,
untouched, uncorrupted by any limitation.
It discovers the boundlessness of Love
that rejects no one.
For love knows no opposite.

Silence is the light of the world,
and it is within you.
Silence is complete unto itself.
In Silence, expression finds its perfection.

Whenever anything has come to fulfillment,
it must touch Silence,
for Silence is fulfillment.

Out of Silence emerges all creation.
If you could come to Silence,
you would see how all the forces focus in you.
The universe blesses you.

See how the planet becomes still.
Everywhere there is the eternal action of Life
forever giving.
Be part of that action
and you will find your own abundance.

Get acquainted with the night sky,
with the fresh breeze of the dawn.
Dawn and twilight exist for man to empty himself,
to know the endless quiet
and the gift of peace which never ends.

We have never seen the kiss of dawn upon the earth.
Every leaf, every twig rejoices in its vitality.
It is the joy of all that exists.
The whole earth awakens with the kiss of dawn,
anointed by the dew.
What a glorious and beautiful world we live in!

Your heart sings songs of joy and freedom.

Dawn bestows a blessing upon creation.
Knowing that, the still mind responds.
Dawn goes away.
But the blessing the still mind has given remains.

Attend to this state of sensitivity.
It becomes the very order of the universe.

All God created imparts peace.

Observe nature as night comes.
A gathering and focus comes to be in nature and in man.
You begin to see the Intelligence of Existence
when you are in a State of Being.
The State of Being has much to give,
and nothing to want.

We need the serenity of twilight to BE.
To BE is natural.
We do not have to seek it.
To BE is accessible if you have the space within
and the time to be quiet.
In the state of BE-ing, time ceases.

Have faith in your own Identity —
the glory of who you are.

Come to the freshness of Silence
and it will be the dawn of your life.
Out of that freshness you will light the world
with the peace within yourself that never seeks.
It is the ever perfect.
It is the source of Life.

Part V

The Voice

THY KINGDOM COME

There is no answer to the Voice for God
Except His Word.
There is no sound except the Voice for God
that can be heard.
For this His Son has ears; to hear God's Will,
And let the ego's voice at last be still.

from *The Gifts of God*
by Helen Schucman

The Voice

From my young age
a silent yearning persisted in me
to meet with men and women
of virtuous lives, of Eternal Voice.

The yearning spoke without words:
a longing for the Voice
that is forever true and absolute,
never caught in indecision,
nor bound to circumstances;
an intrinsic life without opposite,
a law unto itself,
thus consistent at all levels of one's being.

I would perceive the clarity
of such a Voice in a Being,
and the light of its integrity surrounding him,
even more real than the visible form of the body —
the radiance that holds you safe
in its compassion.

I was drawn to be in the atmosphere
of a liberated Voice,
that brings illusions to truth,
and brings man to his own work.
This is its function.

Those who have their own work to do,
such as Nehru, Emerson, Lincoln, and Gandhi,
will not do another man's job.

Not men of causes,
but truly objective,
the few who can think
with words as clear as light.

The highest of these, of course,
is the man whose Voice is consistent
with the Will of God.

And then there are the Extensions,
those untouched by words,
uninvolved in the issues of insecurity,
hence, having nothing to defend:
Lord Rama, Buddha, Jesus, Mr. J. Krishnamurti,
and the author of *A Course in Miracles*,
who rose to the State of the Course, met It,
and shared with us the Thoughts of God It contains.

Those who are holy
impart holiness.
Theirs is the Voice
that shares Thoughts of God.

Theirs is the Voice
with which you cannot argue.
It dispells words,
undoes limitations,
imparts to you your own stillness
and divine function in the world.

Those beings who have something to give
do not commercialize their Voice;
their lives are productive.
They are by nature sensitive to falsehood
and to the illusions of ideals,
appreciative of goodness and lofty attitude.
They are not professionals, men of worldly institutions.

The Voice

A Voice is too majestic to fit into concepts of society.

Where is the immaculate Voice
in a world governed by memory and habit,
shaped by newspapers
and by politicians and professionals,
selling cheap indulgences of self-destruction
and the violence of oncoming disasters?

The gurus of the populace
are men of interpretations
and not of the Voice.

Men of true morality,
having a chaste Voice without contradictions,
affect a change not only at the time level,
but in men internally.

We know so little of this infinite Voice
and its virtuous language of peace and grace.

I had not heard my own Voice
until I was in my early twenties,
and then only for an instant.
The power of it shook my artifical values —
the fallacies of accumulating learning
that trusts only fear.

Yet the unspoken within me was awakened,
and a deep yearning to hear
the unconflicted Voice, consistently true,
became an involuntary force within
that drew me to different parts of the world.

I was an illiterate, naive and too immature,
yet I sought the compassionate Voice
of incorruptible men of virtue.

Those who having no needs,
heed the call of love,
and have the power of God in their Voice.

The virtuous Voice of heroic beings,
those of the purity of Heaven.
Not the heroes of history, the conquerors,
whose self-centered roots were in the earth.

Even in my young age
I disliked organized beliefs
and systems of religious dogma;
yet in my blood
the voices of prejudices instilled in me
were still alive.

How the echoes of other people's contradictions
victimize us all life long
and remain unresolved in most individuals.

There were times when the Holy Word
was almost lost in me,
buried under the words piled upon me.

We suffer the needless depressions
of the imposed errors of irresponsible voices
upon our vulnerable childhood.

The preoccupation of survival
and the sensation of its escape
intensifies body consciousness.
Thus, it limits existence to self-centeredness,
deprived of one's natural state of constant gratefulness.

Observe how men are trained to work,
to sustain and to cater to the pronouncement
of these values.

They exile themselves
into painful loneliness and isolation.
Defending insecurity
becomes the prime preoccupation of their unawareness.

How seldom one ever meets a human being
with a Voice of his own,
who is alive to the Present
and makes the instant holy with his utterance.

The Voice that precedes thought
brings the past into the present,
where the past does not endure an instant.
It extends the present to Eternity.

Any person with a passion for truth
must go through the experience of outgrowing—
undoing the concepts, the conclusions,
and the belief systems of mankind.

Factual men and women
are those who relate with the actual.
They speak what is true
and their Voice brings ignorance to holiness.

For a time, I believed that perhaps out of success
(not knowing what success was)
the real expression of my "calling" might emerge.
But the Voice that precedes thought
would not allow me to get settled
or involved in success.
Thus, the process of outgrowing began.

What becomes of the greatest value
to the one who outgrows
is the Holy Instant.
The moments of intense clarity,

the involuntary moments of Truth
that end duality and make the externals irrelevant.

It is not always easy,
this life of non-conformity and non-external direction.
It fits no category of profession
and one longs to be settled into some vocation.

I was without any external direction,
trying to fit into the externals.
Uncertainty was aggravating to me and confusing to others.
When my preferences were for a bit of safety,
that is what I valued.

Divine Protection
becomes a curse, a torment,
against one's ignorance and resistance.
The ignorance subjects the Eternal Being,
created in the Image of God,
to the tyranny of time and human destiny,
the destructive path of self-centeredness
versus the Divine function of man on earth.

Of course, at the time, one did not know it
as Divine Protection, the inevitable Life Action
that protected me from the personalized activity
of settling down.
The awareness of the falseness came
and spoiled any involvement.
How sacred is true productivity
that sees the false as the false.

Internally and externally,
we, the children of limitation,
are born into a survival consciousness,
fenced in with devices of unreality,
unhappy because of our foolish desires.

What a detrimental environment for growth!

Holiness is not imparted
to the young of age — the innocent.
Manmade laws and values are the basis of education
and constitute an awesome force.
Convention assumes importance
and not the awakening of rebirth,
the awareness of one's eternal Voice —
the God-given Identity.

But how precious is the Voice
that is a Light to an age
caught in the commotion of its interpretation.

I wondered what sacredness must surround
the Voice of Socrates and Lao Tzu,
untouched by deception.
It is rare to find the Voice, everlasting,
amidst men beguiled by their activity of unfulfillment
and the crisis of projected fears.

Where is the space for peace
that takes the terror and the pain away?
Where is the discrimination
of the Voice that abides in every man?

Without the Voice of Love
that undoes separation,
there will be nothing left in the world
but sensation and the cruelty of ambition.

In my case, I met Mr. J. Krishnamurti, who,
in twenty minutes,
had me staggering out of the graveyard
of the meaninglessness.
His words were lifegiving:

The Voice That Precedes Thought

"There are no problems apart from the mind."

You learn as you go along
that everybody wants Truth,
but when the attachments are challenged,
the scene changes.

A Course in Miracles makes it clear
that what the ego wants is what it hates.
Thus, it is inevitable many would choose
not to give up the indulgences of self-improvement
or let the dependencies fall away.
Out of the many, few are earnest.

Not unless you end the words in yourself
can you listen or hear,
for only to a still mind is eternity revealed.

The Voice that precedes thought is always quiet
for it speaks of peace.
It is only as loud as your willingness to hear.

The Voice you hear is but your own.
It is given to you by God
Who asks only that you listen.
His is the Voice that calls you back
to who you are.
It says you are the light of the world.
Awaken, and be glad!

BOOK TWO

Part I

Gifts from the Retreat

Introduction

These are excerpts from "Going Beyond Problems To The Discovery Of One's Own Inner Calling — A Ten Day Retreat" held in Santa Barbara, California, August 10-19, 1980. Ninety people gathered from all parts of the country.

The action had begun with the weekend workshops given in different parts of the country. Each workshop imparted the benediction of gratefulness. The group which gathered for the Ten Day Retreat emerged from these sharings.

The benediction that we are now beginning
to discover is the ACTION OF LIFE —
involuntary and impersonal.

Action is ever of Life.
Every moment its perfection is complete.
It is independent.

Action deals with the individual
and not abstract ideas.
It deals with the goodness in each
and awakens one to higher values.
Circumstances and the assumptions of consequences
it dissolves.

These excerpts are meant
to renew, revive, and share
the spirit of a Retreat
which went beyond knowledge
and where the transforming, Divine energies
surrounded the individual.

The sessions always started
with one hour of silence.
What Divine forces it invoked!
Then a song, sung by Pamela.

The main speaker was not a man of words —
but the wordless blessing
of which we all became a part.

The laughter, the prayer, the profundity,
the intense moments of silence and
the benediction of the gaps within the thought,
led to the discovery
that man is of the Kingdom of God.
Some even flowered beyond silence.

The Retreat is still alive.
The action continues to unfold
its clear direction.

Man has a fascination for expansion.
It is in proportion to his sense of littleness —
but only as long as he is oblivious
of his own boundlessness and his holiness.
The discovery of our boundlessness and
perfection is where we are headed.

The atmosphere was such
that out of the ninety people present,
eighty signed up for the Forty Days
in the Wilderness Retreat
to take place the following year —
the next step of serious study.

These gifts from the Retreat intimate
the perfection within us,
and direct us to:

Love ye one another.

Gifts from the Retreat

I hope you came never to go back.

There is only one place
where one comes to —
the living moment of the present.

There is no other place.

* * *

In the present, there is no past.
When you are caught in the past and future
you are without the energy of the living moment.

You are whole only in the moment.
Then you can listen
and your mind becomes receptive
to the Thought of God.

IN THE PRESENT THERE ARE NO PROBLEMS.

All problems are of the mind.
All insecurity and fear are manmade.
Unless you are free of the past
you will never see beyond appearance.
Unless you are in the present
your eyes are sightless.

In the continuous present

the fruit ripens, the river flows, it rains,
and there is the circulation of the blood.
It cannot circulate yesterday or tomorrow —
it can only circulate now.
There is only the present.
And that is continuous.

When you are not in the present
you start projecting — a wish, separation,
a torment for yourself;
ignorant of perfection
you are barren of the vitality
of the living moment.

The living moment is cleansing.
It is the only fact.
The rest is opinion and assumption.
If you are not with the moment
whatever you say has no validity to it,
no integrity to it.

THE PRESENT MOMENT ALLOWS NO DECEPTION.

For us to know the action of God,
we have to come to the living moment.
That is the only reality.

FEAR CANNOT LIVE
IN THE REALITY OF THE PRESENT.

THE PERFECTION OF THIS MOMENT
IS A BENEDICTION OF GOD TO US.

Silence
Stillness

THE PURPOSE OF EVERYTHING
IS TO BRING THE MIND TO STILLNESS.

Spend time each day in quiet.

Put aside the preoccupations of the day.

You are much more with nature
when you are quiet inside.

The still mind is ever present,
thus transforming energies of Heaven
accompany you wherever you go.

A silent mind is the Mind of God —
free of all mortal fear, belief, and knowings.

Why do you evade the stillness?

NO KNOWING CAN BE WORDED.
KNOWING IS A STATE OF STILLNESS.

ONLY IN STILLNESS IS ONE WHOLE.

All mental thought, all activity is partial.

Partial awareness is caught in deception,
in activity that is personal.

In total awareness you become whole
and your body becomes the temple.

In silence it becomes so content —
as if annointed with holiness.
That relates you with all that is.

When the mind is still,
it is very energetic.

When you fear,
when you dislike,
your mind can never be still.

The ambitious mind can never be still.
It has sacrificed stillness and
sacredness for restlessness.

When you are really quiet,
then you realize
that there is no such thing as time.

When you are in a state of truth,
you are timeless.

Silence is receptive to the Grace of God —
belief is not.
Belief makes man fanatic, divides.
Belief does not know the truth.

SILENCE CLEANSES DECEPTION.

Silence is ever there
like the Grace of God.

The only purpose of words
is to come to silence.

Only a silent mind can perceive
what a fact is.

Only a silent mind can live a fact.

Knowledge does not apply to reality.
Silence does.

Truth
Timelessness

THE ONLY TRUTH IS *WHAT IS.*

Only in truth duality ends.

In thought, concept, dogma, belief systems,
there is always conflict.

IF YOU CANNOT HARNESS THE ENERGY OF A TRUTH,
YOU ONLY KNOW THE WORDS.

When you and the truth become one
you are productive,
of Heaven,
not of the earth,
and you bring to the earth
the love of Heaven.
And that makes you a co-creator
with God Himself.

Any moment you want to
you can discover your timelessness,
your divinity.

We speak that we do know,
and testify that we have seen. . .
John 3:11

EVERYTHING IN LIFE IS TO BRING YOU
TO YOUR TIMELESSNESS.

All that is of time will perish.

That which is eternal,
nothing of time can touch it.

A seed is a gift of timelessness.

NOTHING REAL CAN BE THREATENED.
NOTHING UNREAL EXISTS.[1]

Action of Life

NO SEEKING
JUST BEING.

In Being
you discover
what your real purpose in life is —
not in doing,
not in activity.

Activity is never productive.
The Action of Life always
frees one of activity.
Let us become the Action of Life,
free from the past and future
and therefore the co-creator.

It has a newness
that continues with every breath.

Once we are not related,
we are lonely and unfulfilled,
and then we get caught in activity.

Activity is ever devoid of action.
Activity is personal.
Action is of Life.
Action is never helpless.

The Action of Life is to free man
from deception —

from his getting away
from the living moment.

Our reality is not of *how,*
but of the action within.

When you are with this living moment,
your life is part of the Action of Life.
It is not separate.
It begins to manifest the Will of God.

A NEED AND A MEETING OF A NEED
IS ONE MIRACLE —
it is the Action of Life.

How — The Means

THERE ARE NO MEANS TO IT.
IT IS.

How is ever the false premise,
which makes you dependent on something else.

The means and the *how* produce the deception
and the illusion.

Without the *how,*
you would be in the present.

HOW IS HELPLESSNESS.

As long as we live by words,
we continue the ordeal of the *how.*
Every time you think and plan,
there is a *how.*

The origin of activity is a deception.
The origin of projection is a deception.
How makes us helpless.
How violates our perfection.
How is the seed of all illusions.
Where there is seeking, there is *how.*
Where there is projection, there is illusion.

Helplessness

Most people are preoccupied
with their helplessness,
which is not a reality.

We are in crisis
because we believe in helplessness.

To discover what God has made you
is to outgrow all helplessness.

Wisdom begins with KNOW THYSELF.

It is an inner awakening,
and you discover yourself entangled —
helpless.

Helplessness is always caught
in a situation.

Every person who is not
with the Will of God
is caught in a situation.

A situation is a doing
and we have to bring activity,
which is a doing, to an end.

HOW IS HELPLESSNESS.

Wisdom

Wisdom begins
with the discovery of who and what you are.
Your brother is a mirror to show you
how you act towards him
is what you really are.

A Problem

THERE ARE NO PROBLEMS.
PROBLEMS ARE CREATED BY THE MIND.

We need discrimination
to bring deception to an end.
The minute you have choice and preference,
you are in duality.

ANYTHING YOU CAN LOSE IS AN ILLUSION.

We are in crisis
because we believe in effort.

Fact

Your words have no meaning
when you are not with a fact.

All facts are born out of silence,
humility, innocence —
stillness.

When you come to a fact
the separation ends.
The illusion is gone.

THE LIVING MOMENT IS THE ONLY FACT.
IN THE PRESENT THERE IS NO FEAR,
NO WORRY, NO INSECURITY.

A fact is the discovery of something that is true.
A fact is that which is lived each day.

To be with a fact
is to be free from all your knowledge.
Your mind is still;
it is in its correct place.

All of humanity learns and is enriched
when one human being learns a fact.
The fact carries with it the blessing
that we share.

WHAT WOULD THIS WORLD BE
IF EACH ONE OF US
HAD A FACT TO IMPART?

There would be no wars.
All men would be affected.
When one being discovers a fact,
the whole of humanity —
every consciousness —
is affected by it.
Because human consciousness is
related to the earth's crust,
when we change our consciousness
the crust of the earth is affected.

Knowing a fact is to be free —
free of all conclusions,
all concepts, all belief systems.
Once you are free
you are in the present,
and the present is the only fact.

You cannot interpret a fact.

That which ends duality is
not subject to interpretation.

ONLY A FACT CAN CHANGE US.

The fact will free you.

THOUGHT CAN NEVER KNOW A FACT.

The fact is spiritual.
Conclusions, assumptions, opinions are not.

The fact has Divine energy in it.
It is at once clarity.

No helplessness ever touches a fact.

We are here to come upon fact
and be energized.
In the same instant you receive it,
it is to give.

You will receive it
when you feel deeply
that the other needs it.

You do not give anything
that you have taken from another.

When you give a fact and a truth to another,
it is for all life,
for the eternity of his being.

When you give it, you receive it
for it is the expression of love.
This receiving is from Heaven.

Thought

WITHOUT THOUGHT THERE IS NO TIME.

The purpose of thought
is ever to bring one to the clarity
of the living moment,
to bring the mind to stillness.

That which is timeless continues.

When you are clear,
there is no thought.

THOUGHT CAN NEVER KNOW A FACT.

When we are preoccupied and defeated,
ever caught in thought,
we cannot even receive a blessing.
But when you are whole,
you are no longer of thought.

The Mind

The mind
will never come to stillness
until you have stopped dissipating
the energy in things unessential.

THE MIND IS WHAT YOU PUT INTO IT.

The wise never puts anything
into his mind.

Projections

When there is conflict
you cannot live without projecting
wishes and wanting
because you are dissatisfied.
You are not related.

When your mind projects a future
you have fear,
and you never realize
the perfection of this moment.

Awareness

NEEDS ARE MET BY AWARENESS.
WISHES ARE DISSOLVED BY AWARENESS.

Awareness is productive
because awareness dispels deception.

If you are attached to this or to that,
you will fight awareness
because you have vested interest.

Awareness brings you to the correct place within
and time stops.

The discovery is
that there is nothing to want —
it is all perfect.

> *WHEN THE ARCHER MISSES THE BULL'S EYE,*
> *HE ALWAYS FINDS THE CAUSE OF THE ERROR*
> *IN HIMSELF.*
>
> Confucius

Success

You only seek success because
you are not in the present.
To want success is
to discover, "I am insecure."

Renunciation

RENUNCIATION IS RENUNCIATION OF FEAR,
NOT OF THINGS.

Productivity

You receive Divine energy
when you are productive.

Silence and productivity
go hand-in-hand.
Those who withdraw
do not know what real silence is.
It is too energetic.
It is not passive.

To be in silence is to know productivity —
for it makes man the co-creator.

Silence gives you the vitality
to be productive in the real sense,
not with unessential preoccupation and activity,
but with the inner unfolding
where you express the beauty
of your own eternity.

Right Living

SIMPLIFY.

KEEP THE DISCONTENT BURNING.

Find your own inner strength
so you can walk alone,
and by walking alone
be related to all that is —
just as the stars in Heaven
are related to all that is.

When you live right,
that rightness is a breath of Heaven
upon the earth.

DO NOT SELL YOURSELF.

DO NOT LEAVE THE TIMELESS.

BE STRONG.

CAN YOU LIVE A LIFE WHERE YOU WILL NEVER
TAKE ADVANTAGE OF ANOTHER?

Without that you will never learn
what it is to live in relationship.

Relationship is at the level of love.
Where there is advantage
there is no love.

Where there is love
you stand alone,
but you are related.
Thinking can never bring you
to non-conflict.

Like Noah, you must build an ark.

Your ark is your own life —
the purity you bring within your own life
of eternal light.
Nothing can destroy it.

Be very responsible
to say nothing you do not mean.

All you have to do is to be true
to one thing in your life,
and it will bring order into everything else.
It will transform you.

We have no idea what the efficiency of devotion is.
We know the efficiency of fear
and the efficiency of greed.
Wait till you have the efficiency of devotion —
you will be purified.

Learn to put away the non-essentials
to merit that kind of perception.

Knowing nothing knows all,
for it is in touch with reality.

One right action in your life
is a seed you have sown.

"I may not know what is real,
But I will not go for the false."

When you are independent of the external,
you cannot be deceived by yourself
or another.

Every pleasure carries,
inherent in it, sorrow.
That is a law, a fact.
There is no pleasure without sorrow.

Unless you find the joy
and happiness within you,
you will be subject
to the pleasure and sorrow
that comes from the external.

ONLY THAT WHICH IS LIVED IS THE TRUTH.

That which is born of the flesh is flesh;
and that which is born of the Spirit is spirit.
John 3:6

Purpose

Man is endowed to be the co-creator
when he is in his correct place within.

Intent is prior to thought.
When you have the intent,
all things open before you.
It makes all things possible.

Self-assertion only begins
when we do not know what our real purpose is.

WE ONLY HAVE ONE INNER CALLING.

Our purpose is Divine.
The action is of Love.

Self

FREEDOM IS FREEDOM FROM THE SELF.
It is not something another
bestows on you.

Freedom is freedom from your own
projections and efforts.

Once you become concerned about yourself,
you limit yourself
and make yourself small.

Freedom is at the beginning
and not at the end.
J. Krishnamurti

IT CAN BE BUT MYSELF I CRUCIFY. [2]

Never to Blame Another

FROM THIS DAY
YOU NEED NEVER BLAME ANOTHER.

We need to put away so many preoccupations.
We know so much.
But we do not know application.
Will you learn never to blame another? —
For that is ingratitude.
Will you learn not to be afraid of consequences? —
For then you will have conviction.
The fear of consequences is only
because you are attached.

Why do you dislike people?
Because you are afraid of losing something.
You never see your poverty,
but you blame the other.

How out of place is blame!

JUDGMENT IS OF SEPARATION.

Not to blame another
would free you from
the temptation of ingratitude.

When you do not blame another
you will also learn the miracle
of harmlessness.

When you are harmless
only then you are peaceful
and pain will not touch
a man who is peaceful.

Then you can resolve and confront
your own poison, your own hate.

Perfection

Once your eyes are open,
you see the perfection in everything.
You are never put off
by the idiosyncracy of another person
for you do not relate at that level,
you relate in a state of love
in which both become one.

YOU ARE TIMELESS.
YOU ARE NOT OF THE EARTH.

Do not ever underestimate yourself,
for that is a denial of your own glory.
Nothing exists in the universe
that is not to serve you.

All the benedictions of Heaven are upon
the person who wants to discover
the perfection within himself.

Ye have not chosen me,
but I have chosen you,
and ordained you,
that ye should go
and bring forth fruit,
and that your fruit should remain:
that whatsoever ye shall ask
of the Father in my name,
he may give it you.
John 15:16

You must question things of time and thought
and really see they are all illusion

born out of your desperate need to be somebody
and to deny you are already perfect.

A new dawn begins within you
when this activity comes to an end.
Then you sing songs of joy
you have discovered within —
you are independent of all things external.
You discover that you are not separate from anything,
the reality of yourself.

When you have fear or loneliness,
you violate your own law.
For you are a law unto yourself.

What is timeless needs no belief.
It is ever perfect.

There is a place
so sacred and impeccable within us
that is kept intact by the Love of God.

That holiness,
that sacredness is ever there
in you and me, in everyone,
so that at any instant you can discover
your holiness, your eternity, your wholeness.

We seek no goals, no success, no projects, no plans,
only the perfection of our being.
Without that,
we will never know what our real purpose is
for being on this earth plane.

The perfection of God's creation
is never complete
until you have perfected yourself.

THE DIVINE PLAN REMAINS INCOMPLETE
UNTIL YOU AND I HAVE PERFECTED OURSELVES.

There is work to do,
the work of being co-creator.

You are a co-creator when you love another.
Your conflict has ended;
you are no longer in duality.
Then you can love.

We have to come to God's perspective
to know who we are.

THE PURPOSE OF EVERYTHING IN THE UNIVERSE
IS TO BRING THE SON OF GOD
TO THE ACTION OF PERFECTION.

Everything that is created is complete.

Each child is a saviour of the world
with the whole of creation in support.

Divine Grace comes to liberate
the children of God from illusion.

The action of Grace is always to bring you
to your real nature.
The action of thought and ignorance
is always achieving something outward.

EXCEPT A MAN BE BORN
OF WATER AND OF THE SPIRIT,
HE CANNOT ENTER THE KINGDOM OF GOD.
John 3:5

Relaxation

RELAXATION IS
THE EFFECT OF TIMELESSNESS
ON TIME BOUND THOUGHT.

If we do not have relaxation,
we externalize life
and destroy ourselves.

Relaxation brings timelessness to thought.

Relaxation, wisdom, and simplicity
are the same thing.

RELAXATION BRINGS US TO THE HARMONY
OF *AS IS.*

Discipline

Discipline is not something
you impose upon yourself.

Discipline is not to deceive.

It invokes awareness.

It is natural to love one another.
We just have to remove deception
through discipline.

Learning

THE ONLY THING TO LEARN IS TO GIVE UP ILLUSION.

The man who is fully awake
is always with the learning
and never with the past —
the knowledge.

A student has to give life for life.

Learning is to *Be* the Children of God.

Listening

LISTENING IS FREE FROM INTERPRETATION.

To listen is to be without
the activity of the brain.

When you are whole,
then you listen.

You can only be whole
when you are no longer of thought.

The act of listening will introduce you
to your inner calling.

Urgency

Everything that is significant
and important carries with it urgency.

Urgency comes into being when you begin to see
what is important in the real sense —
what is basic,
what is profound.

EITHER YOU AVOID THE CRISIS OR YOU ACT.

The Will

In order to come to discrimination,
we discover the will.
The will has no images.
It does not seek.

THE WILL OF GOD
IS ALWAYS IN THE PRESENT.

The Individual

ALL THINGS MAY FAIL,
BUT THE INDIVIDUAL WILL NOT FAIL.

Yoga

YOGA IS THE HIGHEST MORALITY.

It is a way of life where you and God
hold hands together.

It cannot be commercialized.

It can only be shared one-to-one
with those who have lost
the fear of consequences,
with those who want to come to their own eternity,
for it is timeless.

Relationship with God

If you are in right relationship with God,
you have no problem.
If that relationship is not established,
there is no end to problems and disharmony.

The first principle is to right that relationship.

WE NEED TO GET THE ACTION OF THE SPIRIT
ALIVE IN US.

The life of the Spirit is different.
It is not demonstrative.
It has no beliefs in it.
It is of Heaven.
It is eternal.
It is not personality stuff.

EVERY SINGLE VOICE AND SOUND
THAT TAKES PLACE ON THIS PLANET
IS A CRY FOR GOD.

ALL THINGS OF BEAUTY ARE THERE
TO REMIND US OF GOD.

The Lord came
and we recognized Him not.
But He left without leaving.

Christ is a State.
In that State no one is divided
and everyone is a Son of God.

The Man of God

The mystic sees
something beyond the beauty

— THE HOLINESS —

and he lets it be.
He owns nothing,
for everything that is, is sacred.

All men of God free people
from their superstition,
from their preoccupation.

The man of God never seeks success,
never organizes.

HE JUST IS.

So there comes a man of timelessness
to the level of time, to affect it
and to free man from his "isms,"
that he may have the compassion
to love the God within the man
and not his rituals.

THE MAN OF GOD AWAKENS. HE DOES NOT TEACH.

You go to a teacher
to die to deception.

A true teacher has nothing to teach.
He introduces you to your own perfection,
and frees you from your own deception.

Gratefulness

GRATEFULNESS IS THE KEY
THAT OPENS THE HEART.

The Grace of God inspires your heart
with a sense of gratefulness.

We must bring our consciousness
to a state of gratefulness.

Gratefulness opens one,
purifies the consciousness,
changes our whole attitude.
It brings about a sensitivity
and an awareness
that we are not fully utilizing.

THE ONE TEMPTATION IN THE WHOLE WORLD
IS INGRATITUDE.

Gratefulness is when there is laughter
within every cell of your being.

A beggar's hunger consumes the mind
that is separated from God —
always want, want, want.
How can it be grateful?

To be liberated from wanting
is to know what gratefulness is.

Can gratefulness ever stop?
It is ever continuous.

FOR WHAT YOU ARE GRATEFUL
YOU WILL NEVER BE DENIED.

GRATEFULNESS BRINGS YOU TO WHOLENESS.

THE PURPOSE OF GRATEFULNESS
IS TO SILENCE THE MIND.

Is there ever a moment when gratefulness
can leave you?

When you become that aware,
then you would not need thought
to be grateful,
then your stillness will make
all things sacred.

A man of peace who wants nothing
has no plans, no projects.
He brings you to the God within
and makes you what you are.
He does not make you dependent.
He makes you aware of your own glory.

When you come to that state —

I can of mine own self do nothing. . .
John 5:30

you have completed
the discovery of gratefulness.

Love

Love never has an enemy.
It is never afraid.
Love cannot blame anyone
for loss and gain have no meaning.

To know is to know holiness.
To know is to see holiness.
To know is to impart love.

The timeless action is always of love.
Any insecurity or fear is out of place.

There is no need of fear.
It is self-made.

When *you* have nothing to give,
only then begins the giving
because only then you can give of self.

In giving the brother the love,
you receive the love of Heaven.
Find out what prevents you
from loving one another.

The Apostle Peter only received
the gift of God when he wanted to give it.

You cannot receive it
until you want to give it
because it is of God.
You cannot manufacture it.

Only the gift that you give becomes yours.
In giving it, you receive it;
in taking it, the other receives it.
And that is relationship.
All the rest is dependence.

You can only receive love
when you have it to give.
Therefore, the man of God
is ever productive.
He is giving clarity to every life.

God loves His children,
and He only has one child.
Love is His child.

His child is a State of Being,
not a physical vessel.
You are His child when you are not separated
from another.

Then you have no one to blame,
no one to cheat, no one to react to.
You meet at the level where you are both One.
Then you will know what wholeness is.
You are not in conflict with anything.
Your life becomes creative and effortless.

You know love because you are productive.

LOVE IS PERFECT, NOT THOUGHT.

Thought is never clear.

LOVE IS THE ONLY REALITY.

The Lord bless thee, and keep thee:
The Lord make his face shine upon thee,
and be gracious unto thee,
The Lord lift up his countenance upon thee,
and give thee peace.

Numbers 6:24-26

An Individual's Experience

The following letter was selected from many written by the Retreat participants.

* * *

August 21, 1980

Teacher, Friend and Brother:

Thank you. There are songs of joy singing in my heart — songs that are not of words. There is dancing here also. Perhaps it's the Soul rejoicing. There is poetry being recited here. It feels like a joyous celebration is going on in my heart. I can't begin to describe it — only feel it.

The skills of written communication have eluded me. So I'll try to convey with this unskilled ability my gratitude for a beautiful and wonder-filled retreat. I am grateful for the opportunity and facilities provided and all the beautiful and blessed people who graced me with their presence.

On August 10 through August 19, 1980, each day there were assembled in the Chapel of La Casa de Maria in Santa Barbara, California, ninety people who had come from afar and near.

I am one who had come for an inner awakening into the *Action of Life* where there are no problems. Each of the several sessions was preceded by a period of Silence. From being silent I became acutely aware of many and different "sounds" coming from the surroundings. As I listened to these sounds, I noticed that my body became calm and still. As I brought my attention within, I became very much aware of the chatter and dialogue going on in my mind. I became disturbed because this never ceasing dialogue wouldn't let my mind be at peace.

Speaking with Tara Singh about this problem on Wednesday, the fourth day of the Retreat, gave me clearer insight. He asked me if I would try to get rid of a waterfall. To which I readily responded, "No, it's natural." Then he said: "It's not natural to get rid of a part of oneself."

I hadn't realized that I was trying to get rid of something that is part of me. I only recognized I became highly frustrated from trying to still my mind. He told me to not make it a problem and leave the fighting to the marines.

I am grateful that he pointed out this fact to me. Each time thereafter, when I sat still to observe the Silence, I became aware of the inner dialogue and began to notice what was taking place, what was happening in me.

Well, I was amazed to realize how frequently I am absent from the Present. I was surprised to find out how much my mind was involved in activities arising from the memory of past events and experiences, my projections of ideas, fantasies of future probabilities, in "maybes" and "what ifs."

Between this chatter and dialogue were *tiny* gaps. Only in this small space I became aware of the Present. I realized a living vitality, an energy, a Silence that stilled the mind and calmed every cell in my body — a Presence that filled my being with Light (it felt like light). At that moment there was no fatigue. All within me seemed to be well and at peace — something indescribable — perhaps a glimpse or fleeting moment of bliss. It seemed to encompass my whole being. In that moment I found myself in a new world — a world where I long to stay and never return.

So the chatter continues — as I watch it I am learning something about myself. I notice that some ideas and beliefs immediately come to an end and others continue. I want to know, why don't all these ideas come to an end?

After I committed myself to the Forty Days in the Wilderness Retreat, the dialogue began: "What if this happens? What if that doesn't work? What if I can't, etc., etc." It was really working me over, pulling me apart. Then suddenly I looked at what was happening.

All these "what ifs" were ideas — non-existing reality — which I was projecting into the future. These ideas were not facts. They were false. They had no life and no energy. I felt burdened. I was absent from the Present. When I became aware of this non-factual projection, they vanished and were gone.

The wisdom was profound. As his austere presence jolted our indifferent and sleeping minds, a gap widened between the chatter; the past and future fled; we were immediately face-to-face with the Present. Wisdom and Love entered our hearts and minds — awakening us — embracing us — teaching us.

Tara Singh pointed out who we are and our purpose for coming to earth. He taught us Facts and how to discern a Fact. (Facts of God and Facts of the Son of God.) He showed us what it is to be in the Living Present and brought us to that state where the living waters flow. He introduced us to our Timeless State, communing with our hearts and minds, reactivating, with love, the "Seed of God" each of us is.

It was during the midday on August 19th, 1980, a few minutes past the high noon hour, it happened — in a chapel, on a hill surrounded by barren and majestic mountains, trees full of fruits, green leaves and blossoms — one could hear the singing of birds, the roar from a plane passing overhead, and noise from the chatter and laughter of people outside the chapel. Everyone was initiated, one-by-one, with the Name of God. Then he turned around, facing the large circle of brethren, and called: "Will you come and bless me?"

So the Retreat was adjourned for nine months and ended with a song of praise.

Affectionately,
Mary Titus

Part II

Holding Hands With You

Holding Hands With You

THIS IS THE FIRST SHARING.

WE WILL CALL IT:

"HOLDING HANDS."*

The idea of a newsletter
had occurred to us long ago,
but we wanted to keep everything simple.

Believe me,
simplicity is the difficult thing in life.
It demands the wisdom and the vision
that goes with "Know thyself."

When we become aware of Divine Laws,
the externals become superfluous
although we may not realize it.

A person interested in the simplicity
of "Know thyself" invariably outgrows
so much of what we call knowledge.
For knowledge of things external
does not apply to Reality.

---

*Between 1981 and 1983, there were eight *Holding Hands With You* pamphlets sent to all participants of Tara Singh's workshops and retreats. They were a means of keeping the one-to-one relationship alive. This is the first *Holding Hands With You* in the series. (Editor)

Mental preoccupations have assumed
such an importance in our everyday life
that we tend to believe
they can replace wisdom, love and God.

Why then would we start a newsletter?

I hope you realize
that we have given serious thought to this.
That means that our purpose is not success-oriented.

I have a dread of organization,
and even more so of taking
other people's money and energy
under the name of communities or ashrams.

Men who have not known
the positive energy of Love
resort to the illusions of projects.

But YOU are important,
not projects and goals,
for these are abstract.
You are the Child of God,
and it is for you this creation exists.
Without your perfection it is not complete.

YOU ARE HERE ON BEHALF OF GOD.

We pray that it would bring you
to the awareness that Self-Reliance
has become a necessity in this externalized age.

For those who are serious,
the simplicity of "Know thyself"
is fundamental to our very survival.

The issue is the discovery
of the perfection within.
The rightness of this eliminates
projections and interpretations
and brings one to the clear direction.

> IN CRISIS
> YOUR OWN INNER STRENGTH
> IS WHAT WILL CARE FOR ANOTHER.

When we are inspired,
the newsletter will be sent to you.

It is meant only for those
who have taken the workshops
and are preparing inwardly
for the Life of the Spirit
and to be as one here in behalf of God.

We look forward to seeing how it unfolds.
This newsletter commemorates
some of the things I have realized
about my function in life.

It is becoming increasingly clear
that I am to work, one-to-one,
with those with whom
Life has brought me into contact
and who are serious.

It is a relationship for life.

I have committed myself
and pray at the beginning
and the ending of the day
for each one who has taken a workshop.

LOVE YE ONE ANOTHER

is beginning to unfold its greater miracle.

This newsletter is to each one of you
sharing my innermost feelings.
What a blessing to hold hands;
by it I am rewarded.
It reminds me of Mother Teresa:

For me, everyone is an individual —
I can give my whole heart
to that person,
for that moment,
in an exchange of love.

We have realized the workshops, retreats,
and the Forty Days are the preparatory part
of a Life Action.

Now we discover that
it matters little what anyone of us
is doing at the present,
or what kinds of jobs we have,
or where we are.

Sadness is the sign
that you would play another part,
instead of what has been assigned to you by God.[1]

Productivity from the Divine perspective
is the elimination of self-deception.

God's perspective is something we do not know.
It will not fit into any personalized existence anyway.

Gratefulness is swifter than words.

Listen to the words of Jesus to Pilate:

Thou couldest have no power at all against me,
except it were given thee from above. . .
John 19:11

I hope you are beginning to see
the wider spectrum of my relationship with each person,
and perhaps the very purpose of man on earth.

It was some time ago
that I began to get intimations about it;
but the realization is only now
beginning to manifest itself.
Thus, it imparts its newness and vitality.

Clarity, certainty, integrity, truth —
all these are aspects of the One energetic Love.

My purpose is to bring those
who are brought into contact with me
to the realization of the one Commandment
that Our Lord gave twenty centuries ago:

A new commandment I give unto you,

That ye love one another;
as I have loved you,
that ye also love one another.
By this shall all men know
that ye are my disciples,
if ye have love one to another.
John 13:34-35

Brothers and sisters,
do you not see that love is the only solution
to our economic, political, marital, and social affairs?

Love makes one independent of all situations.

Behold the significance of this,
and the basic change it makes in our life.

Man has fragmented everything and assumes
that a married person cannot live and be perfect;
or a businessman is not religious;
or carpentry is only a job.

Hence, the thoughts that regulate our existence
are born out of our separation from God.

The tragedy is that we have replaced
God's Thought with our thought,
and perverted our values.
Our very life is caught in duality and conflict.

So, I am not doing workshops as "my thing"
and each person who participates
in them is important to me.

It is you with whom I have the relationship.
I do not identify you with the activity
you associate yourself with.
For it is you Life has brought me in contact with.

One function shared by separate minds
unites them in one purpose,
for each one of them is equally
essential to them all.[2]

We all have the God-given name,
and we all have our Divine purpose.
It is revealed only when we discover the miracle of:

Love ye one another.

Now we see how self-convinced
we are of our littleness.
The truth is that as a Child of God
your bigness does not fit into this world.

There is goodness in each one.
We have to work with that goodness.
The direction is unfolding.

My relationship with you is with your Reality,
not with your judgment of yourself.

I want to share with you also
what came to me recently:

When we have no alternative,
only then can we discover
the miraculous reality of HIS COMMANDMENT:

LOVE YE ONE ANOTHER.

Behold how much joyous energy the Truth imparts,
and the coming to clarity of one's purpose.

God's Will for you
is perfect happiness.[3]

So laugh your laughters.
Make nothing a problem, for there are none.
Your joyousness would uplift another.

I am blessed to know you.
This one-to-one relationship is
the continuous workshop.
It delights me, for it is real.
It is a statement of love and strength.

The workshops are a benediction.
There is an urgency that moves
with the God-given energy
to share the compassion and the gladness
of the heart.

What a discovery —

THY WILL BE DONE IN EARTH,
AS IT IS IN HEAVEN.
Matthew 6:10

I am confident that each one of us can make it.

Here is an excerpt from my Gratefulness Journal:*

"No matter how beguiled or perverted,
or what the world has done to a person,
if there is one being on this planet
who loves that person,
that person can make it.

For this I am grateful.

I am fast finding the truth,
the strength, and the confidence of it.

For this I am grateful.

Thus, it is no longer determined
how bad a person is,
because for me, bad is not bad.

---

*In his workshops and retreats, Mr. Singh encouraged each participant to keep a Gratefulness Journal in which they were to write down things for which they were grateful. This was to be done before going to bed in the evening. (Editor)

Bad is not of Heaven.

My love only acknowledges
that which is of God.
No unreality is its Truth.

Gratefulness is swifter than thought.

Love is free of judgment.
Love is free of weakness.

And nothing is impossible,
for love is not of thought."

LOVE YE ONE ANOTHER

is the miraculous energy of Grace.
The teachers come in His behalf to help people
realize the Commandment He gave.

We will work together,
awakening the laughter within us,
and filling this world with gladness.

The certain are perfectly calm,
because they are not in doubt.
They do not raise questions,
because nothing questionable enters their minds.
This holds them in perfect serenity,
because this is what they share,
knowing what they are.[4]

My dear,
this letter is an unfoldment
of what is evolving in me.

It really clarifies,
better than I have ever known it,
the destiny of my life.

You are very much a part of it.
Our relationship is beyond words.

Stay close to gratefulness.
It is the benediction.

Write your journal with the purity of innocence,
so that it is not a routine or duty
but a flowering in you
in which you are grateful to your Creator
for all that is,
for all that you perceive
and all that you give
in the joy of meeting another's need,
bringing you closer to your compassion's creative energy.

Let it be a prayer.

For what you are grateful
you will never be denied.

We are so preoccupied.
I hear the Lord saying:

"YOU HAVE NO SPACE FOR ME."

Silence is essential.

God's peace and joy are yours.

OUR GIFT TO YOU
IS THE BENEDICTION OF GRATEFULNESS —
THE FLOWER OF THE WORKSHOPS.

Part III

The Preparation

DEDICATED TO THE ANONYMOUS TEACHER OF GOD,
THE PURITY OF A SAINT,
UNTOUCHED BY WORDS.

Introduction

This is the second in the series of

HOLDING HANDS

and represents the preparations for the upcoming Forty Days
in the Wilderness Retreat Experience,
September 1 to October 12, 1981.

It is amazing that in this day and age,
with all the time and costs it involves,
that over one hundred people from all over the country
would come to attend it.

It is an historical event in the sense
that people are coming on the strength
of the one-to-one, individual relationship they have.
Thus it is not towards an abstract project or cause.

We know each person by name.
You are not lost in the bigness of numbers
nor the falsity of success.
There is the bond and the contact.

Instead of encouraging people to come
or even promoting the event,
we have been selective and
we have discouraged many friends from coming.
To some, asking them to reconsider their motivation;
to others, to realize the seriousness of the undertaking.

Many of you in a family situation
who wanted to come were directed towards
the purity of harmony in the family
rather than to come to the Forty Days.

But all and everyone who has taken
the workshops and retreats are included.
The twenty-four hour prayer vigil
will include a prayer for you
every hour of the day and night
in the spirit of:

> *The Lord bless thee, and keep thee:*
> *The Lord make his face shine upon thee,*
> *and be gracious unto thee:*
> *The Lord lift up his countenance*
> *upon thee, and give thee peace.*
>
> Numbers 6:24-26

We honor those whom Life brought us in contact with
and consider this relationship with you a blessing.

It makes us feel good that in spirit you will be with us.
Please pray and unite with us each morning and evening.

As I have said,
after the Forty Days,
I go into the life of contemplation,
having seen the action through to the end —
workshops, retreats, and the Forty Days.
What a joy to see the loveliness of this completion!

I have always admired the Essenes
for their action of disbanding the order
when their job was done.
They became anonymous and free of attachment.

One who is in the process of outgrowing is always ahead.

Solitude is the purity of a saint.
He is truly productive
for he is the extension of God on earth.
Nothing external regulates his life.

Now that I have found the prosperity of abundance within,
set in motion by the higher forces,
I will spend the next period in contemplation.
But I will meet with those of serious intent.
Together we will explore
the reality of being the co-creator
and how we can serve Christ best
and thus move on to the development of a lifestyle
in itself compatible with the work
to maintain optimal levels of energy.

The contemplative lifestyle evolves
from the comprehension
that steps out of the lure of external expansions
with their personality involvements.

Let us now consider the attributes of a student.
The awareness of the student and the teacher
is the highest state of integrity
and has laid all limits by.

Of the teacher-student relationship,
God takes care of the needs.
Both are provided for.
They are in contact with the Source within,
and may become those who are sent with:

. . .neither purse, nor scrip. . .
Luke 10:4

to proclaim:

. . .the kingdom of God is at hand. . .
Mark 1:15

Attention is the student.
Attention gives us the energy
and becomes the actual prayer.
It discovers the Kingdom of God on earth
and the joyousness of all Creation.
Attention is born out of the internal stillness.

Stillness is of eternity.
It is no longer the preoccupation of time
with its exhaustion and conflict
of which doubt and violence are born.

Contemplative life in action is productive.
There will be the exploring and learning
with the serious students.
The energies have changed
and the vitality of the newness
reverberates in me already.

The stillness of contemplation
is the civilizing factor in the world.

It is my joy that I have an obligation
with each one of you who have taken the workshops
to keep the relationship alive.
Occasionally there will be the contact
and the holding hands with you
where you may attend a retreat for the renewal.

You will not be overlooked,
for I value the relationship with each one of you.

My life becomes yours even more.
I consider this my benediction.

ALL THAT I GIVE IS GIVEN TO MYSELF.
THE HELP I NEED TO LEARN THAT THIS IS TRUE
IS WITH ME NOW. AND I WILL TRUST IN HIM.[1]

I am grateful to hold hands with you
and for the pure energy it endows us with.

You who are beloved of God are wholly blessed.

My love to you who inspire me with compassion.
Now I pray that we continue to journey together,
and that you too may keep the relationship alive.

The Preparation

Seek ye first the kingdom of God,
and his righteousness;
and all (these). . .things shall be added unto you.
Matthew 6:33

The past is over. It can touch me not.

Unless the past is over in my mind,
the real world must escape my sight.
For I am really looking nowhere;
seeing but what is not there.
How can I then perceive the world forgiveness offers?

This the past was made to hide,
for this the world that can be looked on only now.
It has no past.
For what can be forgiven but the past,
and if it is forgiven it is gone.

Father, let me not look upon a past that is not there.
For You have offered me Your Own replacement,
in a present world the past has left untouched
and free of sin.
Here is the end of guilt.
And here am I made ready for your final step.
Shall I demand that You wait longer
for Your Son to find the loveliness You planned to be
the end of all his dreams and all his pain?[2]

The preparation for the Forty Days in the Wilderness starts from the moment you receive this letter.

We need to recognize the Divine Forces
that brought this holy instant about.
We need to acknowledge the forces
in our lives that draw us to it.

The right perspective will bring us to reverence —
a purity of feeling beyond the verbal.

Will you then deal with the essentials?

From this day the preparation begins
by weeding out the unessentials in your life.
So that when you come to the Forty Days
on September 1st, you have already
dealt with some of the basic issues.

The issues are:
to bring loose ends and involvements to an end
and bring your life to a new order
in which there is the space to receive the vertical
that awakens Divine faculties.

All unfinished business is to be dealt with
so that a CHANGE can Be.

Unless you do this,
you will go back to the same situation
and be drawn back into the same responses,
urges, impulses, and vibrations.
This is what prevents change.

The energy that can bring loose ends to a head
is free to receive.

Our life must bear the attributes of God.
Spiritually, man is of pure energy.

We are stepping away from the abstract,
from the unclear motivations and getting to the factual.
It is at the factual level where the change is possible.

The Forty Days is a holy event.

Will you come to the totality of action
and discover the potentials of pure energy?

Half-measures merely promote
the inadequacy of postponement.

You have made a decision within you
to come to the Forty Days and
to live a life that is consistent
with the Divine Laws of Creation:

> the laws of love,
> gratefulness,
> and forgiveness.

We are here on behalf of God.

An undisciplined life drags one down
to the sublevels of indulgences —
waste of time, energy, and purpose.

We start with awareness of WHAT IS.
Without any condemnation of self or
judgment of another.

A conclusion is not without judgment.
The knowing of judgment violates
the laws of freedom and harmony.

Forgiveness is not learned.
One does not have to learn the fact.

Ideas may provide more alternatives
but they have nothing to do with reality.

LOVE is independent,
for it is the real — a fact.

That which is independent of ideas
is the clarity of pure energy.
Ideas have no validity.

Points of view and opinions
can argue — the commotion —
but cannot come to the strength of clarity
that is independent of you and idea.

Circumstances and situations that consume
our energies and regulate our lives
are to control us no more.
That is the challenge.

Then there is the challenge of being
in rhythm with distinctly different energies within.
And being in rhythm with dawn, day, and twilight
to improve the quality of blood and sleep.

Without discipline,
nothing of value or order in life is possible.
But discipline in its reality is not imposed,
whereas routine is.

Discipline in its origin is learning.

Routine blocks the function
of different levels of energies
and violates the law of wholeness.

To change to inner direction and attunement:

— begin with the responsibility
of lessening your dependence
on the externals.

— bring order in your life
by seeing things through to the end
and honoring your words,
and inherit the joy of bringing
something to perfection.

— start with paying your debts,
if any, so that you are under
no one's obligation.

— please waste no energy, money,
or anything, and adhere to the
Laws of Conservation.

The Life of the Spirit is the virtuous way of life.
It has its own values and ethics.

Please question everything
that you say, do, and give your energy to.
It is important to be with RIGHTNESS.

But tell me, said Socrates to his friend,
do you really believe that you understand
the ruling of the divine law,
and what makes actions pious and impious,
so accurately, that in the circumstances that you describe
you have no misgivings?

Socrates

Before coming to the Forty Days

we need to air the mind, the heart,
the house and the clothing.
Anything that is unkempt is to be
brought to the order of the essential.

Paint or clean your house and closets.
Make it the external temple.
Either give away things or have a garage sale.
Discover the space of simplicity.

Begin to relate with the good in the other.
Wrongs and rights are secondary.
Discover the goodness in yourself.

If you do not relate with the goodness
in the other, they will not have
the strength of their goodness to cope.

> *Therefore if you bring your gift to the altar, and there rememberest that thy brother hath aught against thee; Leave there thy gift before the altar, and go thy way; first be reconciled to thy brother, and then come and offer thy gift.*
>
> Matthew 5:23-24

It is important that this be done
so that when you return to your abode renewed,
it is consistent with the happiness
and the quality of your life.

We are to be free of the old problem-ridden ways
of much talk and activity that
keeps us in the prison of self-centeredness.

As long as there is boredom,
one cannot step out of indulgences.

We are to step out of being busy.

Productivity is consistent with the Action of Life.
It is not an activity.

Productivity eliminates unfulfillment.

I question teaching and learning.
The sages were never externally educated.
They were never interested in other people's money.

We did not learn to breathe,
to grow our teeth or to blink our eyes.
That which is beyond the ability of words
is the Unteachable.

The action of Grace is there
as much as this moment of reality
is the present.

The man who is at peace has found his own goodness,
and he is not afraid of the Unknown,
nor of making mistakes.

His peace and goodness take care of it.

The wise introduces us to our own perfection.
He establishes nothing.
The Kingdom of God is already established.

The wise does not believe in organizations and projects.
He is the man who is Present and responds to the need.
He has no alternatives.
His life is intrinsic and direct.

Lord Buddha left a kingdom.
Jesus never had a home.
Mr. J. Krishnamurti lived out of suitcases.

You have to master not taking advantage of another
and not blaming another,
for these promote an attitude of dependence.
Helplessness seeks gratification
and will never know what relationship is.

LOVE YE ONE ANOTHER.

What, in-depth, does this mean?
It intimates to us that Love is the law.
Without its pure energy of harmony,
relationship is not possible.

Pure energy comes into being
when we are in harmony with one another.

It is important that we know
the definition of humility.
I wonder if we comprehend
the Divine faculties of its pure energy?

Humility stands independent
and has its own resources.

My brother, peace and joy I offer you,
That I may have God's peace and joy as mine.[3]

Your intent in coming to the Forty Days
must be that you want to eliminate
all that prevents you from coming to:

Love ye one another.

A friend of mine confided to me that:

> "You could not teach or speak of forgiveness and freedom from guilt if you had not gone through it internally in your own life."

We are embarking upon transformation within and without.
It is the action of integrity.

Our needs have their voices,
and we must learn to expose ourselves
and dare to fathom the deep motivations.

Questioning has the vitality of rightness.
The question is blessed and is the prayer.
The true question does not accept the verbal answer.
It silences the mind and brings one to a new sensitivity.

The event of the Forty Days in itself
is John the Baptist, the Harbinger.

Such an event becomes the center of
radial activity of influence in the earth.

We participants are blessed and must not forget
that the preparation begins with
the spirit of gratefulness.

When you are grateful,
you are one with God.

GRATEFULNESS IS AN ECSTASY OF GRACE.
ITS WHOLENESS ELIMINATES INSECURITY.

INGRATITUDE IS THE ONLY TEMPTATION.
IT IS NOT OF PURE ENERGY.

Inner gladness is the prosperity of gratitude within.
Remember, forgiveness is the greatest of all virtues.
It eliminates the past and the internal impurities.

Make sure that before you come to the Forty Days,
you have forgiven everyone including yourself.

Forgiveness offers everything I want.
Today I have accepted this as true.
Today I have received the gift of God.[4]

Only in man the two forces
of Heaven and earth combine.

The transformation is to come to
a life of rightness.
In times of crisis,
it is your inner strength
that will care for another.

Rightness is related to Heaven.
It knows no fear, no insecurity.
Wishings and wantings are not factual.
Opinions and assumptions have no reality.

The blessed stability of Trust cannot be
as long as it relies on the fictitious future.
The past too is unreal.

When man is perfected on the time bound planet,
subject to evolution,
he realizes his own eternity.

Forgiveness is beyond learning.
We have yet to discover the truth of this.

Prepare yourself.
Best that you be in good health.
Go for walks.
Go early to bed.
Rise early and do some exercise.
Wear clean clothes,
and try to stay with natural fabrics.
Fresh food is preferred.

> *An early morning walk is a blessing for the whole day. . .*
> *Truly our greatest blessings are very cheap. . .*
>
> Thoreau

And he stressed:

> *Beware of all enterprises that require new clothes.*

Read *A Course In Miracles* daily.
Step out of time.
Invoke the Presence and slowly
read the lesson as prescribed.

The curriculum is well-specified
in the introduction.
Do not interfere.
Be in line with IT.

Prepare yourself to:

> *BE STILL AND KNOW*
> *THAT I AM GOD.*
>
> Psalms 46:10

We have to conserve the nervous energy
and come to relaxation.
Be with the Heaven within.
The time and space you give *A Course In Miracles*
is the barometer of your growth.

Rescue yourself from the exhaustion of this age.
From this day on, step out of pressures and stress.
Keep yourself calm, relaxed, and spacious during the day.

There is enough time for the ESSENTIAL.

What is to take place during the Forty Days

merits a perceptive attentiveness.
We meet to awaken the subtler intelligence within us.

Please do not underestimate yourself.
The forces that brought the event about
are the same as those that bring you to it.

The Forty Days is a one-of-a-kind experience
in the history of the New World.
The holiness of the event is so overwhelming.
Leave your problems behind.
Be dedicated to the sanctified atmosphere.
It is not a ritual.
It is a very sacred gathering.

It will be required that you do not divert
another person's attention.
We go there not to look at each other,
but to look in the same direction — God.

Reverence is in proportion to your awareness
of your divinity within.
We will be in this state of reverence
with every breath.
Please understand that
we come to silence ourselves.

We will have a twenty-four hour prayer vigil
for peace on earth and goodwill to man
and invoke the spirit of:

Thou shalt love the Lord
thy God with all thy heart,
and with all thy soul,
and with all thy strength,
and with all thy mind;
and thy neighbor as thyself.
Luke 10:27

The preparation includes writing
the Gratefulness Journal daily.

Gratefulness is the awareness
of the Divine Grace at work.
This awareness is the light of a glad heart.
It ends duality.

I want you to bring your Gratefulness Journal
for us to go over,
along with your daily record of meditation
and the time you spent morning and evening
with *A Course In Miracles.*

As pointed out before,
the first thing you do upon waking is
to make note of which nostril is flowing
and what the pulse count is.
Start doing this if you have not done it before.

We need to get to know the individual that you are
and the forces that regulate your life,
the impurities within,
and the quality of your blood.

We in the family are setting
one day a week aside for silence,
and suggest you do also.

The Forty Days is the plow that digs deep
and uproots all that is unessential.

The atmosphere of the Forty Days —
what a cleansing!
What a preparation in itself to be
where so many are gathered in His Name,
in His Presence.

MAKE STRAIGHT THE WAY OF THE LORD.
John 1:23

Ye have not chosen me,
but I have chosen you,
and ordained you,
that ye should go and bring forth fruit,
and that your fruit should remain:
that whatsoever ye shall ask of the Father in my name,
he may give it to you.

These things I command you,
that ye love one another.
John 15:16

Your peace surrounds me, Father.
Where I go, Your peace goes there with me.
It sheds its light on everyone I meet.
I bring it to the desolate and lonely and afraid.
I give Your peace to those who suffer pain,
or grieve for loss, or think they are bereft of hope and happiness.
Send them to me, my Father.
Let me bring Your peace with me.
For I would save Your Son, as is Your Will,
that I may come to recognize my Self.[5]

Part IV

Excerpts from the Forty Days in the Wilderness

Foreword

The Forty Days in the Wilderness Retreat took place September 1 to October 12, 1981, in the Rocky Mountains of Colorado. Around a spacious lake in the midst of pines and golden aspens, over one hundred people, some with their children, gathered to bring their life to a new order and meaning. They came having completed a preparation by attending a weekend workshop and a nine-day retreat with Tara Singh. The workshop and the retreat prepared them to look within themselves and helped to open their hearts to gratefulness. They came from New York, Connecticut, North Carolina, New Mexico, California, Nevada, Oregon, Alaska, and Europe. Their ages ranged from fifteen to seventy years.

In this book we have compiled some of the seeds of truth shared and explored. For each sentence that is written here, an immense background was given by Tara Singh during the retreat. Step-by-step he explored a basic fact until the actual state of consciousness of that fact was shared by the participants. Days of silence were provided between the days of sessions to allow the space for inner application and taking it a step further.

The Forty Days in the Wilderness for Tara Singh was seeing an action through. From the weekend workshops to the retreats and finally to the Forty Days, a clear direction was shared, bringing to the vision of the participants the serious challenges that lie ahead in real spiritual growth and living a life of service. The underlying action of it all was the formation of a non-commercialized, one-to-one relationship, a friendship for life in which two people look in the same direction — towards God.

Tara Singh feels that the Name of God cannot be commercialized. When life is externalized and gets down to the mass level, the one essential thing that is missing is the individual relationship. The so-called teacher has to outgrow his own unfulfillment and ambition for success. Then the purity of the direct contact that brings one to relationship can occur. The wise is ever accessible, never too busy nor important. To him life is sacred and the individual is important. Such a being attracts the earnest and they come, ". . .not to learn but to BE."*

Out of the Forty Days will flower one-to-one relationships. It will not be a planned format but will originate out of the response of individuals to the challenges the Forty Days made clearer. How meaningful it is to be part of an action that is not pre-determined but will unfold from the spirit of each individual.

Joy Kealey
Editor of the First Edition

---

*Refers to the song, "The Seventy," which was often sung during the Retreat. See page 282.

Introduction

You would like to know your purpose on this planet? You would like to know the truth of love — the Reality of God?

Tell me, in what way does your university education help you in this? And in what way does your science and politics aid you in the discovery of holiness? Is not the basis of education and government, with its industrial economy, largely the promotion of self-survival? Is not self-survival still the dominant issue of man and animal on this planet? Has not your consciousness accepted this as law? But it may not be true.

Do you not live by the law of your wishes and not by the law of love? How far humanity is from this! According to you, freedom then is only a fallacy and salvation some servile submission.

"Freedom is freedom from self," said Thoreau. What does this mean to you? Can we continue to ignore this statement by the foremost of the Forefathers? What does it imply?

It is the statement of a law. And Thoreau hands us the Key, as does Abraham Lincoln when he declares:

"RIGHT MAKES MIGHT."

Have we heeded his vertical words? He also said that if this nation stayed with Rightness, it would be given the instruments of safety. And no nation of Europe or Asia could invade its shores, nor any internal strife affect it.

What is it in you that keeps ignoring the given clarity? Would you question? What guarantee is there that you would heed it now?

THOU SHALT NOT KILL.
Exodus 20:13

LOVE YE ONE ANOTHER.

What has any one of us done about this? What would you do now? Would you question your helplessness? Has the incentive for goodness dried up?

Where is the gratefulness that inspires man and brings him to the action of the soul — the indefatigable spirit of creation? Do you not see that man is tired upon this world, without the energy of love and his contact with God? Would you not like to be free from the preoccupation of fear and insecurity and the torment of conflict within and without? In what way has all your accumulation of external knowledge helped you to know your own perfection and the Kingdom of God?

Why do you comply and accept that you are a child of problems? Why do you impose this authority over yourself? Is this survival concept the authentic evaluation of yourself? Are you at peace with concepts? Are you at peace with the external seekings, following the images and illusions of past generations?

The Forty Days poses these questions and gives us the space to confront and outgrow all the bondages of unreality we impose upon ourselves. Would you now come to urgency and seriousness? Our very life is at stake. Surely we have dwelt sufficiently on the destiny of the New World that provides the individual with the initiative to step out of collective consciousness — the world of man's conditioned brain — and the limitation of the physical senses.

The Forty Days, I assure you, is not for the casual approach. We must truly comprehend the truth of:

The power of decision is my own.[1]

in order to realize the benediction of:

I am not a body.
I am free.
For I am still as God created me.[2]

Would you like to come to the intensity of a dialogue? A dialogue is religious for it draws the PRESENCE of ". . .where two or more gather in HIS name."

I have offered you my friendship for a lifetime — a friendship that is a non-commercialized, one-to-one relationship, that is independent of all external factors and not subject to any circumstances. It is free of the duality of likes and dislikes, wrong and right, good and bad. This friendship and the one-to-one relationship are meant for us to come to a dialogue and go beyond thought to an all-knowing, impeccable State where:

. . .I AM STILL AS GOD CREATED ME.

Do you not want to know the truth of:

BY GRACE I LIVE.
BY GRACE I AM RELEASED?[3]

How would you respond to these questions and challenges?

Tara Singh

Excerpts from the Forty Days in the Wilderness

THE FORTY DAYS IS OF THE SPIRIT OF
JOHN THE BAPTIST.

In the Forty Days we could come to APPLICATION.
Its reality is to affect the planet.

Are you the individual who can step out
of collective consciousness?
Will you give all your energy to not deviating?

The Forty Days is to come to
an impeccable quality within oneself.
The focal point is to deal with our resistances
and to change our lifestyle.

The Forty Days is meant to dissolve the image
we project on the other.

THE WORLD OF THOUGHT IS BEING DISMANTLED.

The purpose of the Forty Days is to come to

LOVE YE ONE ANOTHER.

This is a state that does not react.
Let us end this dream
of conflict and separation that breeds fear.
Wrong and right are irrelevant.

We are here to remove the "but" and the "if,"
for we dedicate our life to serve our brother,
and to give of ourself —
thus no longer sell but serve.

The Forty Days is to bring us to:
non-dependency
and self-sufficiency.

The Foundation for Life Action
is to awaken the Divine Faculties
within you.

Citizens of a Meaningless World

Are we citizens of a meaningless world
run by thought?

MY MEANINGLESS THOUGHTS ARE SHOWING ME
A MEANINGLESS WORLD.[4]

Whatever we look at,
we look at through thought.
Body senses — brain thought —
know ambition and fear,
for they only know unfulfillment.
Thought divides things.
Time is created by thought.

You and I have chosen to question
the world of thought.
To come to rebirth
is to know the power of negative thinking.

Can you be so clear
that you will not accept anything of thought?
One moment of truth disassembles
everything based on thought.

Thought has no power over Awareness.
Awareness has power over thought.
Awareness does not recognize thought as real.

Where else would the search for God have reality
but in the unreality of thought?

As long as there is pleasure and gratification,
it is of thought and thus, personal.
Having seen something
beyond the meaningless world
would change your lifestyle.

The minute I separate from Reality/God,
fear appears.
Everything I project is built around this fear.
Is not all thought part of collective consciousness —
thus conditioned?
In reality there is only
the one collective brain we share.

Ending thought requires a swift action.
But you cannot come to action
without awareness and relaxation.
We are ruled by the energy of thought.
We must widen the gaps between the thought.
Conserve the energy
and it will bring us to stillness.

We must allow the spirit the space to manifest.
When we stop projecting,
we set our mind free.

On the non-verbal state, thought does not intrude.
Thus there is no conflict.

FATHER, MY MIND IS OPEN TO YOUR THOUGHTS,
AND CLOSED TODAY TO EVERY THOUGHT BUT YOURS.

I RULE MY MIND, AND OFFER IT TO YOU.[5]

When There Is Wanting

When there is a wanting
there is activity.
This is the opposite of stillness.
Once there is a wish or an insecurity
there is duality.

Out of stillness is born
the Action of Life.

The body senses always want the opposites.
Projection has authority over one —
the personal authority of separation —
the ignorance.

The body senses ever want "moreness."
Better the slavery of Egypt
than the torment of "moreness,"
for it forbids the action of Moses.

Silence is not a wanting.

How can you fit the Unknown into the known?
We always come up with something out of the known.
THE WANTING IS OF THE KNOWN.
How disastrous is this limitation.

All we have to see
is that we create the illusion.
In reality there is nothing personal in life
to be done.

Kali Yuga and the Consequences of Affluence*

In this age the lie becomes the truth
and the truth becomes the lie.
The quest for pleasure has degraded everything.
What science discovers,
it tends to exploit;
therefore, there is no reverence.

A society rich in means,
poor in purpose.
E. F. Schumacher

In the system
you learn to compromise.
Do we ever come to disillusion?

The system rules society.
And everyone is a captive to it.
E. F. Schumacher

Irresponsible people becoming religious
make of it a dogma.
When religion becomes a belief system,
it is the tribal approach.
We trust in separation and not in God.

---

*A yuga is an age, or epoch, according to Hinduism. The present age is referred to as Kali Yuga, the Dark Age in which man becomes engrossed in materiality. (Editor)

Kali Yuga, the Dark Age, is:
— preoccupation of self
— identification with the body
— addiction to sensation.

To come to the eternity of one's being
is to end an eternal cycle
of all yugas.

Gratefulness is eternal and ever there.

A Course In Miracles

A COURSE IN MIRACLES
IS THOUGHTS OF GOD.

It is to be lived,
and not just to be read.
It is our lifestyle
that makes this difficult.

If you listen to your own thoughts
you will not change,
for change is not born
out of the duality of thought.

WHAT IS A MIRACLE?

It is a reminder from God,
when heeded.

Inherent in every word of
A Course In Miracles is the blessing
that bestows the miracle,
and the separation ends.

Reading *A Course In Miracles*
is God sharing His Thoughts with you.

If you read *A Course In Miracles* for a year
and did not make it secondary,
you would be out of society.

The next step:
Application of it in relationship with others —
becoming a Teacher of Eternity.

What Is Important

How can anything that is not real be important?
What is in the realm of unreality has no value.

A thousand interpretations make nothing important.
When something is not important, one goes astray.

What is important?
What prevents us from it?

Would you come to the purity of a saint?

We feel that the only thing that has value
is at the activity level.
Activity is mostly promoted by self-interest.
"Doing" is of self-centeredness.

Selfishness is important to thought.

What we usually call "important" enslaves us.
But what is important is complete unto itself.
It cannot create dependence.

FORGIVENESS OFFERS EVERYTHING I WANT.[6]

Do you realize what this means?

It means that you never separate
from the Thought of God,
the action of Love.

In forgiveness is where the split
does not take place.

APPROACH IS IMPORTANT.

To wisdom, "important" means
the vision of that which is eternal.

Experience

What is experience?

The Man of God is not interested in experience.
He is of bliss and light.
Experience is limited to the senses.
To that which precedes the senses,
nothing is external.

The external is a projection.
It does not exist.

Judgment is born of experience.
Until we see God in the other,
we will not be at peace.

Making right use of the senses
can free one from deceptions.
Will you listen?

To be precise
is the right use of senses.
It gives one the space to be.

The purpose of experience is
to bring us to timelessness,
to the soundless movement of Life,
the Timeless.

If we deviate into the "other,"
we have deviated into the physical senses,
for thought is external and always a reaction.

BODY SENSES ARE EXTERNAL.
IT IS THE MIND OF GOD THAT IS REAL
AND NOT THE BODY SENSES.

To see the body as anything
except a means of communication
is to limit your mind
and to hurt yourself.[7]

All wisdom is accessible.
There is a level where time and space do not function.
A state of meditation is not recorded as experience.
What is beyond words is not recorded.
You can only tell IT by its effect.

"Father, I Will Not Judge Your World Today."[8]

REALITY IS EVER PERFECT.
Interpretations are not.
We have not seen the world of God,
only one of thought interpretations.

FATHER, UNLESS I JUDGE, I CANNOT WEEP.[9]

Can one be at peace
listening to one's own interpretations?
Peace is of Reality.

We are not to wish things to be different
from what they are.
For in the moment of wishing
we have isolated and separated ourselves.
This is the origin of conflict,
of duality,
of "me and you."

The moment we are in conflict
we deprive ourselves of the holiness of unity,
of the gift of love.

There is no world.
What we call the world is something external,
and *A Course In Miracles* says:

NOTHING REAL CAN BE THREATENED.
NOTHING UNREAL EXISTS.

If you accept the external world,
you inevitably accept fear,
insecurity, and unfulfillment.

Our thinking is ever inconsistent.
Its truth is a lie.

I loose the world from all I thought it was.[10]

You cannot be at peace
until you are part of the whole.

I am one Self united with my Creator,
At one with every aspect of creation,
And limitless in power and in peace.[11]

When you say *one Self,*
nothing is outside of you.
There is no good or bad anymore.
Everything of the Universe is focused in you.

We are no longer meaningless.
We are boundless.

IN CHRIST THERE IS NOT THE DIVISION,
FOR HE ENCOMPASSES EVERYTHING.

"If I defend myself, I am attacked."

If I defend myself I am attacked.
But in defenselessness I will be strong,
And I will learn what my defenses hide.[12]

Limitless in Peace,
you will never react to anyone.
Reaction takes place within you
the instant you deviate into the personality.
If you are reacting,
you are regulated by thought.

Either you react, which is born of separation,
or you respond as an extension of God.

A Man of God takes all the impurities
of an environment and absorbs them in himself,
dissolves them —
and retains his purity.

The so-called "other"
is to introduce us to the oneness of Love,
the miracle that ends the separation.
Ever be with the miracle of *Love ye one another* —
the benediction of Life.

THOUGHT IS REACTION.
There is always an idea or some "thing" external
for thought to react to.

TO HAVE OTHERS CONFORM TO YOUR OWN IDEAS HOLDS SWAY OVER YOU.

FREEDOM IS FREEDOM FROM SELF.

The action always starts with oneself.

Impurities react.
See what the impurities are in yourself
rather than reacting to another.
You receive GRACE in that very instant —
but first you have to conquer
reaction, anger, and judgment.

From *A Course In Miracles* we learn
the truth of forgiveness
and the freedom of non-judgment,
and realize:

I am not a body.
I am free.
For I am still as God created me.[13]

Grace and looking within go together.
You cannot separate them.
Listen to your brother when he is angry.
Hold his hand and say:
"I am here to serve."

One man in a country
who is capable of "turning the cheek"
has affected that country.
Where is the wisdom in defending separation?

One regrets making another suffer.
Strong measures are always against someone else.

This is never justified
for it is a violation of:

Love Ye One Another

and the spirit of non-judgment.

If another blames you,
give them the space — do not react —
and learn to give Grace in response
rather than retaliation.

Wherever there is reaction, there is effort.
Effort is when you separate yourself from Grace.
It is the illusion of "BECOMING."
Every reaction is calculated,
thus a half-measure.
It cannot give.

Interpretation denies direct response.
Direct response is free of thought.
If you respond directly,
it is never half-measures.
THE DIRECT ACTION OF LOVE AND TRUST IS OF GOD.

Love meets all needs.

Grace is not an experience.
It is a State.
Your action of Grace —
when you do not react —
awakens Grace in the other.

There is so much loneliness and insecurity in the world.
We must make a vow not to add to it,
never to hurt another.

It can be but myself I crucify.[14]

You only "crucify" your reactions.

This is my Eastertime.
And I would keep it holy.
I will not defend myself,
Because the Son of God needs no defense
Against the truth of his reality.[15]

A Dialogue

Religion is a dialogue
between two people, or more,
not based on winning a point.
A dialogue unfolds the newness
and is blessed by the Divine Presence
our intensity of interest draws to it.

Learning is not enough.
What we need is the vitality to discover
the REALITY beyond the words.

The truly religious mind moves
from fact to fact to fact —
thus is free of opinions and assumptions.
The fact is the freedom from the personality.

How thought splits itself and
starts the dialogue within;
and thus, does not amount to anything beyond chatter.

All our problems are born out of our vagueness.

Watch the casual mind —
how it breeds desires
and projects unrealities.

Learn to formulate the right question,
an eternal question of man.

All manmade concepts are threatened by a question.

"What is right action?"
is to find out who is asking the question.

It is the challenge that makes one serious.
The challenge rightly faced
is the beginning of a prayer in you.

When we come to crisis we have the energy.

The truth is neither the question nor the answer.

Intelligence is an energy that demands clarity.
Intelligence is an urgency.

Clarity ends duality and brings peace.

In the absence of peace
we get caught in our interpretations.
But to the silent mind is given
the dialogue with God.

Relationship

What is the relationship between Truth
and the daily life of personality?

Relationship is within the realm of the One
not with the personality.
Personal relationships are based on motives,
and attachment is the source of sorrow in the world.

Sentimentality is born of confusion.

Dependence is always sustained by gratification.
And where there is gratification, there is sorrow.

In self-sufficiency there is relationship.
Nothing in creation exists in isolation.

We come to this plane to bring
the Thoughts of God to earth,
to our brothers and sisters.
Love introduces us to eternity,
to the eternity of the so-called "other."
Thus in love the separation ends.

This is relationship.

Are We Afraid of Freedom?

Fear is the crux of the matter.

We want to come to action and be free,
but there is the fear of consequences.
Thus we never discover
the vitality of action.

When Buddha left his kingdom,
he dissolved the fear of consequences.
CONSEQUENCES CAN ONLY BE PROJECTED.

Mr. J. Krishnamurti dissolved
the worldwide organization
of "The Order of the Star"
which proclaimed him as the World Teacher.
People asked: "How could you do this?"
He answered:
"Without the fear of consequences."

Will fear accept any answer?
It will project another fear.
If the motive of the question is fear,
will it accept anything else?

In actuality what does this prove?
It proves that Truth cannot be learned,
but only realized,
for it is eternal.

Action is always free of consequences.

It is not personal.
Action is not learned.
Action is of Life.
It is involuntary and free of duality.

We do not change
because fear and insecurity are our belief.

Alas!
The bondage of our beliefs is opinion.

We may want to outgrow the external
but we do not realize
the fallacy of our preoccupation with it.

We make everything external,
and then we are afraid of it.
All our activity is to keep the fear intact.

When you come to Reality,
activity ceases.

The basic issue:
How the self continues
and thought promotes it.

By projecting fear,
I separate myself.
Could you see this as a fact?

Fear is our own interpretation.
It is not Real.
What prevents you
from seeing this as a fact?

Wherever there is fear or insecurity,
there is self-centeredness
with its interpretation.

Where there is interpretation,
there is no meeting the Oneness with God.

A meaningless world engenders fear.[16]

Either you are in the realm of love
or you are not.

Love always responds.
Fear always reacts.

The Illusion of Seeking

Are we not trying to come to the present
by growing away from it?

Throughout the ages man has tried
to change from "this to that."
Ideals have been worshipped.
But from "this to that" invents time
and promotes conflict.
Is this not the basis of organized religion?

Truth is a state in which there is
no pursuing or seeking.
Let us not struggle from "this to that."
Let us discover what Grace is.
For there is no conflict in the state of Grace.

We keep the activity of seeking alive,
and remain unaware of Gratefulness.

. . .serenely unaware of everything. . .[17]

What is "everything?"
Is it only the personal preoccupation of
"this to that?"

IS CLARITY A GRADUAL PROCESS?

IS LEARNING INVOLVED IN CLARITY?

Excerpts from the Forty Days in the Wilderness

With all the educational systems and universities
where is the God-lit man freed from his duality?

"Here" is not a place.
It is a state of being.
Nothing happens any place else but in the present.

Are you present?
What prevents it?
Is it not the preoccupation of
"from here to there?"

Whenever the mind moves, it is confused.

> *Disquietude is always vanity.*
>
> St. John of the Cross

Behold a state
of impeccable stillness and silence.

In the THOUGHT OF GOD
there is no "from here to there,"
for the THOUGHT OF GOD is audible
only to the silent mind.

To be an extension of the Thought of God
demands CHANGE.
We deviate into the past.
Devoid of the vitality of the present
we remain ignorant of love.

We become productive only in the present,
for only love is productive.

Behold the contradiction:
Whatever we seek
we never really want to find.

Seeking is always a movement away from truth.
Beware of the fallacy of "from here to there."

The STATE is always present.
But thought deviates,
and through deviation we seek
to get back to the STATE.

To end the separation
you need no activity.
Activity inevitably ends up in habit and routine.
Conditioning has its deep roots in the past.

I see only the past.[18]

Seeing the false as the false
is the action of freedom,
the present on the past.

What Is Our Purpose?

The will has a relationship with our purpose.

Once we acknowledge what we *are*
we will not acknowledge
what we have "become."

The notion of so-called "becoming"
is the world of collective consciousness.
Each person's purpose is to awaken
himself and the other to their REALITY.
"I am going to help the people" is still
the unfulfillment of "becoming."

A purpose is wordless.

There is an action independent of "becoming"
and personality.
You and I can realize our Divine purpose
and end the activity of "becoming."

Here we meet to end the fallacy of "becoming"
and to BE.

> *I am one Self, united with my Creator,*
> *At one with every aspect of creation,*
> *And limitless in power and in peace.*[19]

We are here to discover our wholeness —
the perfection that we are —
and to be the Teachers of God.

There is no Choice

At the level of "me and mine"
no one is free.
All they want is a better set of choices.

What is choice but a separation?
Can deviation into choices know Reality?

IN THE PRESENT THERE ARE NO CHOICES.
THE PRESENT HAS THE VITALITY OF PERFECTION —
THE ETERNAL.

Perfection lives forever
and is independent of time.
It is of the Kingdom of God.

What is time bound is not perfect.
There is no freedom in it.

Fear must have choices to protect itself.
A Course In Miracles tells us:
The minute you make a choice
it is an attack on God.
ALL EFFORTS ARE THE CONTINUATION OF ATTACK.

Thought projects choices.
Choice personalizes life.
Do you not see that at the moment of choice
you isolate yourself?
What else but choice deviates from God?

Gratefulness discovers perfection is already there.
In a non-choice state you are whole and holy —
one with God.

A friend well said:

"I was trying to make choices in a choiceless world."

*Not to "Learn" but to "BE"**

We are addicted to the world
and our daily routine
which makes us intentionally dull,
and to our mania of learning
which teaches us nothing.

Truth is not taught
but only recognized.
External learning is not enough.
We have to CHANGE.

Ideas never know Reality
and are born of thought.
Externally you cannot know Reality.

LEARNING IS A PARASITE THAT LIVES
OFF OUR DIVINE ENERGY.

Words are to be brought to realization.
But it is "safer"
to be with the status quo of non-change.

"There are no problems apart from the mind,"
said Mr. Krishnamurti to me.

The solution is never an activity.
The solution is the awareness of the false.

---

*Refers to the song, "The Seventy." See page 282.

THE NEXT STEP IS:
"NOT TO LEARN, BUT TO BE."

We have to bring the learning to its appointed end:

YOU ARE THE CHRIST.

This is the purpose of the Forty Days.

Can We Change Our Lifestyle?

Simplicity introduces one
to self-sufficiency.
Without simplicity,
you will never know non-dependence.

Will you value
and resort to the expediences of the system
or will you live according to
your own conviction?

Your unwillingness to change your lifestyle
binds you to the world of compromises.
To have something of God to give the world
eliminates insecurity
and frees one from the external system.

The earnest question will emerge
out of the urgency to change the lifestyle,
FOR THEN THE WHOLE LIFE WILL BE
BEHIND THE QUESTION.

IS YOUR LIFESTYLE NOW
RELATED TO REALITY?

Discipline

What ambitions lurk behind your disciplines?

Discipline is an awakening.
It is not a learning from another.
Discipline means learning of self.
It is not an activity.
It is an awareness.

Denial is a waste of energy.
True discipline does not deny anything.

Discipline means not to deviate into choices.

Discipline brings all that is unessential
to an end.

Discipline is the eye that sees
the false as the false.

When you come to discipline,
you are in harmony
with the rhythm of creation.

TO BE THE ACTION OF LIFE
TAKES THE POWER OF DECISION.

THE POWER OF DECISION IS MY OWN.
THIS DAY I WILL ACCEPT MYSELF AS WHAT
MY FATHER'S WILL CREATED ME TO BE.[20]

Seeing

To see the false as the false
is the end of all that is external.
We are self-spent,
for we do not see the false as the false.

"Seeing" is free of deception
and independent of personality.
"Seeing" is always present;
it is not bound to time.
It is the Divine light of eternity,
the holding hands with God.

Do we ever see?
We look but we do not see the whole picture.
Truth cannot fit into partial attention.

The Action of Life is effortless,
for it is of God and not personal.

Gratefulness is the benediction.
Gratefulness brings us to the state
where we see all is provided,
and of strife and efforts we are liberated.

Negative thinking is the highest form of intelligence.
It does not get involved in anything.

There is no attachment in seeing the false as the false.
The "seeing" is the energy that liberates.
There is no greater energy in the world.

Enlightenment is an explosion of the "SEEING."
"SEEING" is the extension of God.

Then you live in the manifest world,
but you affect *it*.
Man is both manifest and the unmanifest.

Silence

On the day of silence
observe what takes place.
That which observes is silent
and is not touched or contaminated.
It is an independent action of a still mind.

TO SILENCE, FREEDOM FROM DECEPTIONS IS ESSENTIAL.

In the purity of silence there is no "you,"
nor is there thought without you.

All things complete themselves in silence.

Silence has no past to be influenced by.

All we have to do
is silence our thought interpretations
to hear the THOUGHTS OF GOD.

Where external thoughts are silenced
that place is consecrated.
Where a man of silent mind dwelleth,
it bears the light of a temple.

THE STILL MIND IS BOUNDLESS.
IT IS BEYOND SPACE AND TIME.

Silence frees us from the body and its senses.
It is not intruded upon by physicality.
Thought is a halfway state.
The body uses thought for the things it craves.

The intense vitality of silence
gives the body peace, purity, and a freshness.

Father, this is Your day.
It is a day in which I would do nothing by myself,
but hear Your Voice in everything I do;
requesting only what You offer me,
accepting only Thoughts You share with me.[21]

To Give Attention to the Unknown

It is attention that discovers the Unknown
and recognizes the truth.

Logic outgrows itself,
because it comes to the Unknown
and does not retreat into belief.
Logic gains in its vitality
and by its keen questions,
thought can be dissolved.

Logic demands precision and refuses deception.

Fact frees one from bondage.
The fact is vast, and ever-continuous.
It is the Action of Life.

Logic comes into being "not by oneself."
The strength of logic
is the energy of attention, absolutely unbiased.
It questions the very movement of thought.
It is the source of stillness.

Silence is too energetic for thought to manipulate it.

Right Action

When we are in the *state* of non-anxiety,
choices end, and we know
what right livelihood and right action are.

RIGHT ACTION HAS NO GOALS.
IT DOES NOT SEEK.
RIGHT ACTION ORIGINATES FROM THE *STATE.*

There is no such thing
as a place or circumstances
to that *state.*

THE LAW OF LIFE IS TO BRING ORDER.

Simplicity has all the Life Forces behind it.
It would know how to bake bread for God.

The *state* is a quality of Being.
Whatever it touches is blessed.
The material world needs
the blessing of the Action of Grace.

ACTION IS ALWAYS AN EXTENSION OF GRACE.

The function of the Life Forces is to meet a need.

LIFE MEETS ALL NEEDS.
If you have understood this,
how could you be touched
by the anxiety of insecurity?

Love Is Not a Sentiment

In relaxation the gap between the thought widens.
Love comes through the gaps between the thought.
When we are externalized,
we cannot receive the love.
Love is not a word.
It is not personal.

What a blessing!
Seeing the brother invokes love
and ends the separation —
my "seeing" him ends my duality.
He gives me the wholeness.

Love is the creative force that sustains life.
It is by Grace we live.

The Will

WILL IS OF GOD.

Will is the action of the REAL,
and we are part of it.
IT is independent of the thought process.

That Will relates to that which is eternal.
The Will needs no education.
The activity of the brain needs education
because it operates at the experience level.

By the Will
we recognize the Truth.
The Grace of the Father
takes the final step,
and gathers the Son unto HIMSELF.

Forgiveness

Forgiveness in its actuality
precedes thought.

The "forgiveness" of thought
is when one tries to forgive.
But true forgiveness is freedom
from the outset.

I trust my brothers,
who are one with me.[22]

Freedom does not accept the world of the body senses as reality.

The body can be a means of
communication of the past,
or it can communicate the present.

Only my condemnation injures me.
Only my forgiveness sets me free.[23]

Forgiveness does not abide in a world that changes,
but brings one to that which is changeless.

Forgiveness offers everything I want.
Today I have accepted this as true.
Today I have received the gifts of God.[24]

WHEN YOU COME TO THE ETERNAL,
YOU WILL SEE THAT IT IS YOUR FORGIVENESS
THAT BROUGHT YOU THERE.

The Memory of God

Touched by the memory of God
WE ARE HERE TO CHANGE THE DESTINY OF STARS.

The memory of God is an impeccable space
intact within each one,
guarded by His Grace.

The remembrance is not verbal.
It is a PRESENCE.
You attain it not through learning.

Every time you love another
it is in the remembrance of God.

Every blade of grass sings of His perfection —
inspires the Presence of God.

The Man of God

No teacher who is a teacher ever converts another.
He acts from his compassion.
He who converts is insecure and unfulfilled.

The action of wisdom
does not tell you what to do.
It just inspires.
It is not a doing.
This liberates one from the authority of another.

A voice is a voice
when it relates to Reality;
therefore it represents eternity.

THE NAME OF GOD CANNOT BE COMMERCIALIZED.

You do not need skills to bring
the Kingdom of God to earth.
The Man of God is an extension of Grace.
HE AFFECTS.

He does not see the "two."
He does not live by "cause and effect."
The cause is always Grace,
for that is the only action.
All the rest is effect.
His action is the ending of fragmentation.

The saint is a child, vulnerable, harmless.
Harmlessness has no fear of consequences.
This is what the world has to learn.

Innocence is harmless.
The child has two parents for protection,
and the mother's milk
in the breast for nourishment.

Innocence is not afraid of those
who manipulate or harm.
Love cannot be manipulated.

Non-attachment cannot be taken advantage of.
Why are we afraid of being innocent?
Can we learn anything when there is fear?

The Man of God is a teacher
who introduces you to your own perfection.
Therefore, he is anonymous.
He imparts the ability to question
so that you recognize
the self-sufficiency of Grace.

What a discovery!

The Lord does His work.

A teacher of God
is anyone who chooses to be one.
His qualifications consist solely in this;
somehow, somewhere
he has made a deliberate choice
in which he did not see his interests
as apart from someone else's.
Once he has done that,
his road is established
and his direction is sure.[25]

Rebirth

To come to rebirth is to see how we avoid it.

What is rebirth?

It is the action of the Holy Spirit
that realizes:

Nothing real can be threatened.
Nothing unreal exists.

Gratefulness is the first sign of rebirth.
The consciousness of the personal prevents rebirth.
Gratefulness ends the isolation of the personal.

I am in need of nothing but the truth.

And for that peace, our Father, we give thanks.
What we denied ourselves You have restored,
and only that is what we really want.[26]

Gratefulness

Gratefulness is an awareness
of the blessing that is ever-flowing.

Realized gratefulness imparts
all the energy we need.

Ingratitude is the only temptation.
Ingratitude is the impurity of unawareness.

In the absence of gratefulness
we seek pleasure,
and where there is pleasure
there is sorrow.

Gratefulness is the joy of all living things.

Gratefulness is a *state.*
It *is* always there.
It soars beyond incidents
and physical experiences.

The Holy Spirit is my only Guide.
He walks with me in love. And I give thanks
To Him for showing me the way to go.[27]

"By Grace I Live. By Grace I am Released."

By grace I live.
By grace I am released.

Grace is an aspect of the Love of God
which is most like the state
prevailing in the unity of truth.
It is the world's most lofty aspiration,
for it leads beyond the world entirely.
It is past learning,
yet the goal of learning,
for grace cannot come until the mind
prepares itself for true acceptance.
Grace becomes inevitable instantly
in those who have prepared a table
where it can be gently laid
and willingly received;
an altar clean and holy for the gift.

Grace is acceptance of the Love of God
within a world of seeming hate and fear.
By grace alone the hate and fear are gone,
for grace presents a state
so opposite to everything the world contains,
that those whose minds are lighted by the gift of grace
can not believe the world of fear is real.

Grace is not learned.
The final step must go beyond all learning.[28]

To know Grace, you have to let go.

AWARENESS IS OF GRACE.

Nothing else can be called awareness.
If you are not aware of Grace,
you are not aware.

Be aware, and you have made the contact.

GOD IS GRACE.

Grace is not learned.
Learning is a preoccupation of "becoming."

In Grace we hold hands with God each day.

Confidence is a moment of perception
that reveals Grace.
Grace releases us from all efforts.

GRACE BRINGS US TO SILENCE,
AND THE ACTION OF GOD BEGINS.
WHAT A REBIRTH.

TO KNOW GRACE PROCLAIMS *GOD IS.*

The action of Grace is to end separation —
the only deception.

HEALING TOO IS THE ENDING OF SEPARATION.
NOTHING ELSE IS HEALING.

Would you give yourself the space
and look at the world
and see it as ONE?

The purity of a saint
never acknowledges separation,
thus ends the duality within of choices.

In spite of us, God is at work.

EFFORTS GONE —
I AM AT PEACE.

Even "you" are not there in the state of Grace,
for the "you" and the "me and mine"
are external to Grace.

Whom could you harm;
of whom would you be afraid
if there were not the other?

By the Grace of God,
when we look externally we discover
our oneness in our brother.

The outer is a manifestation of what you project.
With the dawn of Grace the illusions end
and peace is,
FOR THERE IS ONLY THE ONE NAME OF GOD.

"I Offer You My Friendship"

On the day of the new moon, September 28, 1981, in an atmosphere of spacious silence, fifteen participants at a time entered a room with a long table on which seven blue candles were lit. They joined hands with Tara Singh and invoked the Presence of Christ. For half an hour in prayerful spirit, a bond of lifelong friendship and clear direction was established.

THE SHARING:

THE CONTINUED ONE-TO-ONE ACTION
AS IT UNFOLDS

I honor your coming and acknowledge the Divine energy
that brought us here to the Forty Days in the Wilderness.
An internal action is already at work
in those who are here for the Forty Days.

Consistent with it, here is the second step:
this is the preparation for those
who come not to learn but to BE.
To those who want to come to the energy
of clear direction of purpose,
change of lifestyle,
and who want to give it the first priority —
I offer you my friendship.

For a year live consistent with *A Course In Miracles*,
for the Course is meant to be lived and not read.

Each lesson invites the memory of God to come again,

. . .and would be sufficient for salvation,
if it were learned truly.

WE HAVE A FUNCTION
THAT TRANSCENDS THE WORLD WE SEE.

Thus,

. . .we give these times of quiet to the Teacher
Who instructs in quiet, speaks of peace. . .

and teaches you:

. . .what to do and say and think,
each time you turn to Him.

He will not fail to be available to you,
each time you call to Him to help you.[29]

THE TEACHER OF GOD

End all loose ends and come to your own resources
so as to be non-dependent on others.
If you sense the urgency, and are interested and ready,
after this year, we will share Yoga.

Yoga, the gift of God to man,
is not to be commercialized.
Our purpose here is to do God's work, not our own.

Yoga is for those who undertake to live:

. . .under no laws but God's[30]

and to realize:

The power of decision is my own.[31]

The one-to-one relationship
consists of the intelligence of cooperation
and invokes the energy of pure harmony.

I trust my brothers, who are one with me.[32]

This makes life the productive expression of Gratefulness:

. . .requesting only what You offer me,
accepting only Thoughts You share with me.[33]

To the Man of God nothing external matters.
He has overcome reactions
and realized the truth of forgiveness.
His response is independent,
for it is the extension of love.
His world is beyond the body senses.
Thus he discovers the truth of:

I AM NOT A BODY.
I AM FREE.
FOR I AM STILL AS GOD CREATED ME.

I was told:

"Get healthy and strong.
You will be needed.

The path is clear.
The way will be shown,
and there is Service."

A Course In Miracles prepares you
for the ministry of God.
Your real function is:

MAKE STRAIGHT THE WAY OF THE LORD
John 1:23

BY GRACE I LIVE.
BY GRACE I AM RELEASED.
BY GRACE I GIVE.
BY GRACE I WILL RELEASE.[34]

The Prayer Vigil

On the first day of the Forty Days a room was set aside for a twenty-four hour prayer vigil. Here two or more people every hour sat before the lit altar and prayed — holding hands with humanity. The very first day the atmosphere of the room changed and became so charged with the PRESENCE that people came out with tears in their eyes.

The following is one of several prayer guidelines created.

* * *

PRAYER VIGIL

Sit a moment.
Drop everything.

Start with THE LORD'S PRAYER:

OUR FATHER WHO ART IN HEAVEN,
HALLOWED BE THY NAME.
THY KINGDOM COME. THY WILL BE DONE
IN EARTH, AS IT IS IN HEAVEN.
GIVE US THIS DAY OUR DAILY BREAD.
AND FORGIVE US OUR TRESPASSES,
AS WE FORGIVE THOSE WHO TRESPASS AGAINST US.
AND LEAD US NOT INTO TEMPTATION,
BUT DELIVER US FROM EVIL:

FOR THINE IS THE KINGDOM,
AND THE POWER, AND THE GLORY, FOREVER.
AMEN.

Matthew 6:9-13

FATHER, THIS IS YOUR DAY.
IT IS A DAY IN WHICH I WOULD DO NOTHING BY MYSELF,
BUT HEAR YOUR VOICE IN EVERYTHING I DO;
REQUESTING ONLY WHAT YOU OFFER ME,
ACCEPTING ONLY THOUGHTS YOU SHARE WITH ME.[35]

THIS HOLY INSTANT WOULD I GIVE TO YOU.
BE YOU IN CHARGE. FOR I WOULD FOLLOW YOU,
CERTAIN THAT YOUR DIRECTION GIVES ME PEACE.[36]

Lovingly pray
with a grateful heart.

After a long pause:

INTO CHRIST'S PRESENCE WILL WE ENTER NOW,
SERENELY UNAWARE OF EVERYTHING
EXCEPT HIS SHINING FACE AND PERFECT LOVE.
THE VISION OF HIS FACE WILL STAY WITH YOU.
BUT THERE WILL BE AN INSTANT
WHICH TRANSCENDS ALL VISION,
EVEN THIS, THE HOLIEST.
THIS YOU WILL NEVER TEACH,
FOR YOU ATTAINED IT NOT THROUGH LEARNING.
YET THE VISION SPEAKS OF YOUR REMEMBERANCE
OF WHAT YOU KNEW THAT INSTANT,
AND WILL SURELY KNOW AGAIN.[37]

Come to wholeness.
Read your lesson of *A Course In Miracles.*

Sit quietly for twenty minutes, then offer prayers
for all those who have taken the workshops and retreats
and those who are here at the Forty Days
by His Divine Grace.

THE LORD BLESS THEE, AND KEEP THEE:
THE LORD MAKE HIS FACE SHINE UPON THEE,
AND BE GRACIOUS UNTO THEE:
THE LORD LIFT UP HIS COUNTENANCE UPON THEE,
AND GIVE THEE PEACE.
Numbers 6:24-26

FATHER, I am grateful.
I pray YOUR blessing
be upon THE PEOPLE OF ALL NATIONS OF THE WORLD
and for PEACE ON EARTH AND GOODWILL TOWARDS MEN.

FATHER, I am deeply grateful for *A Course In Miracles* —
the THOUGHTS YOU share with me daily,
as YOU hold my hand.

Strengthen my mind
so that I may be the bearer of YOUR Commandment:

LOVE YE ONE ANOTHER.

Help me, FATHER, to be non-judgmental
and to end the reactions that take place within me.
I pray that YOU bring me to
the limitless peace of forgiveness.

Free me, FATHER, from:

MY MEANINGLESS THOUGHTS
. . .SHOWING ME A MEANINGLESS WORLD.[38]

FATHER, my heart is most grateful.
I pray that YOU awaken the Divine faculties
in each one of us at the Forty Days in the Wilderness
so that we may come to a rebirth in ourselves.

FATHER, I thank YOU for the benediction that:

I AM UNDER NO LAWS BUT GOD'S.[39]

FATHER, we ask YOUR blessing and healing upon:

(the prayer list)

FATHER, I am grateful to YOU.
I pray for all those in the prisons of the world,
that they come to recognize
the impeccable being that they are
and use their time to turn within
to discover their own Holiness.

FATHER, I am grateful.
I pray that YOUR blessing be upon the sick,
and those in the hospitals all over the world.
Endow them, LORD, with the awareness
that sickness is not a reality.
It is we who make sickness real
by giving it validity.
Thus, it is we who can call upon the Will
and be whole.

Heavenly FATHER, we need YOUR help and guidance
to change our purpose and direction on this planet.
FATHER, we now pray most earnestly
for YOUR Grace to turn the human energy
to serving one another
and the impoverished and the hungry in the world.

Free us, oh GOD, from violence
and the self-destructive insanity of the armament race.
Bring us, YOUR children of Divine energy,
to THY HOLY WILL.

FATHER, it delights me that:

I AM ONE SELF, UNITED WITH MY CREATOR,
AT ONE WITH EVERY ASPECT OF CREATION,
AND LIMITLESS IN POWER AND IN PEACE.[40]

Bring me to the Truth, LORD, of:

I AM NOT A BODY. I AM FREE.
FOR I AM STILL AS GOD CREATED ME.[41]

Thank YOU, FATHER, for the "charge:"

WE ARE THE HOLY MESSENGERS OF GOD
WHO SPEAK FOR HIM,
AND CARRYING HIS WORD TO EVERYONE
WHOM HE HAS SENT TO US,
WE LEARN THAT IT IS WRITTEN ON OUR HEARTS.
AND THUS OUR MINDS ARE CHANGED
ABOUT THE AIM FOR WHICH WE CAME,
AND WHICH WE SEEK TO SERVE.[42]

Songs

INTO CHRIST'S PRESENCE

Into Christ's Presence will we enter now,
serenely unaware of everything
except His shining face and perfect Love.

The vision of His face will stay with you,
but there will be an instant
which transcends all vision, even this, the holiest.

This you will never teach,
for you attained it not through learning.

Yet the vision speaks of your remembrance
of what you knew that instant,
and will surely know again.[43]

THE CERTAIN

The certain are perfectly calm,
because they are not in doubt.
They do not raise questions,
because nothing questionable enters their minds.
This holds them in perfect serenity,
because this is what they share,
knowing what they are.[44]

SERVICE

We will serve and do God's work,
and He will provide.

The external is not our concern.
This means freedom from future,
where fear and insecurity abide.

I need not look at the world through thought,
but will serve and do God's work,
and He will provide.

The Name of God cannot be commercialized.
Thus the one-to-one relationship is without fees.

I will be available to serve
and to help bring your loose ends
and involvements to an end.
So that you, too, are free from
"the meaningless thought that sees the meaningless world."
Thus, bring love to earth.

THY WILL BE DONE IN EARTH,
AS IT IS IN HEAVEN.

THE FORTY DAYS

The Forty Days in His Presence
He shines His face upon us.
His Light shines in our hours of silence.
His Light shines amidst our words and in our sleep.

He accompanies us through our walks in the woods,
looks through our eyes at the lake and at you.
So we may learn to love.

The Forty Days, He shines His Face upon us,
holds hands with each one and shares His Thoughts.

The Forty Days in His Presence
lays the foundation in our life for Divine action.
He shines His Face upon us,
holds hands with each one.
Holds hands with you,
and welcomes us to share His Thoughts
and to be in His Eternal Light.

BENEDICTION:
Our Lord shines His Face upon us.
His blessings surround us.

These Forty Days I embrace you with my life.

WISDOM

What is the energy of wisdom?
Wisdom is the dawn of innocence.

It eliminates the false.
Wisdom is untouched by deception and illusion.
It brings the Divine order in life.
Wisdom is free, pure and impersonal.

It is the innocent light
by which we see the perfection of God's world.

The function of wisdom is that it liberates man
from the self-made world and from the earth forces.

The wise man extends the Kingdom of God.
He is of the simplicity of Heaven,
never gets entangled in thought.
Thought is of the world.

Thus the wise remains uncontaminated by the external.

Wisdom is holy and complete unto itself.

Wisdom is the silent purity.
Silence, knowing all, is contained.

Wisdom extinguishes the burning unrealities of thought
and imparts peace.
Thought yearns for the peace of wisdom.

In the absence of wisdom,
thought becomes abstract,
and in the end, self-destructive.

Wherever there is violence, hatred, and fear,
it is the absence of relationship between wisdom and thought.

The pure energy of wisdom is ever harmonious.

The wise always corrects the error with love;
thus invokes the forces of Heaven
and awakens Divine faculties.

Without wisdom there is no virtue.
Righteousness is its proper name.

Wisdom is eternal.
Wisdom is the might.

TWILIGHT

Twilight comes like the wisdom of the age,
having outgrown the commotion of the day,
to the serenity of the evening.

The flaming colors of sunset
greet it with triumph over activity
and the commotion of the day.

The twilight in its atmosphere of blue
dissolves shadows — ends the duality in man.

The solitary man stands vertical,
his head high in Heaven and feet on the ground,
for he is whole, taller than the sun and moon.

The twilight comes with its intensity of blue,
and there is serenity upon the earth.

In relaxation all things are complete at last.
Tensions disappear.

Peace upon the land is the vibration of twilight.
What a benediction the calm holiness of each evening.

In the sky appear the stars
and in the carefree man
a new sensitivity and its appreciation
that surrounds him with gratefulness.

Twilight is an inspiring call
to become one with the rhythm of life and day.

It bestows relaxation and prepares man
for the miracle of sleep.

At night the senses gather themselves.

In sleep all distractions end,
and God caresses His child.
The Father caresses His child,
shares the ecstasy of His moments with you
and makes you Himself.

Again the awareness is renewed that you are He.

As the soft light of dawn spreads upon the land,
you awaken to your awareness
and bless the whole world all day long.

The glow of the strong, energetic day matures.
The day is essential,
and everything in life is productive.

Whatever you see and touch is blessed
till twilight descends
with its benediction upon the land.

THE SEVENTY

There was a school where the Seventy met
to be in the Holy Presence of Jesus.
NOT TO LEARN — BUT TO BE
and to bear the vibration of His touch eternally.

Yes, there was a school
where the Seventy met to be with Jesus.
NOT TO LEARN — BUT TO BE
and to bear the vibration of His eternal touch.

And so with Ramakrishna the apostles met
in the sacred atmosphere of His Presence.
Not to learn —
but to BE of silence.

To the school of the prophets of holy Elijah too
came the children of God
to know the function of their divine energy —
the ministry of God.
NOT TO LEARN — BUT TO BE.

And now the time is approaching
for the teacher to appear,
and for students already emptied of self to meet.
Not to learn but to Be
the focus of planetary forces,
wherever they go, wherever they are.

Will you heed the call?
End your external learning, for you are not a body.
You are free, for you are still as God created you.

Will you heed the call?
End your external learning —
and BE.

Part V

Profile of Our Age

Foreword

The Foundation for Life Action
is a non-profit, educational foundation
not geared to success but to Rightness.

Somehow, we discovered that people go towards success
when they have failed in goodness and gratefulness.

Our objectives are:

> We have established a bond with those
> who have taken the workshops and retreats
> and offer a one-to-one, non-commercial relationship.
>
> We encourage those that Life
> has brought us into contact with
> to discover the perfection in themselves.
> We do not project abstract projects or plans
> with other people's money and energy,
> nor intend to create dependence.
> We believe in the pure energy of harmony
> and gratefulness, and the joy of perfecting
> what we do.

The Name of God cannot be commercialized.

Vision has no cost to anyone.
It can only bless.[1]

Man is sustained by the Love of God.

By grace I live.
By grace I am released.[2]

From the Foundation for Life Action,
we now rise to the individual one-to-one relationship of:

LOVE YE ONE ANOTHER.

Those of you who have taken the workshops and retreats,
with you we have established a timeless bond
independent of circumstances and the externals.
You are our brothers and sisters
and are ever welcome to come and visit with us,
and break bread.

I share God's Will for happiness for me.[3]

All things are echoes of the Voice for God.[4]

Profile of Our Age

We have returned from the Forty Days
exuberant with its blessing.
We now see the need, the urgency and the wisdom for:

A CHANGE OF LIFESTYLE

I am determined
to see things differently.[5]

We have started at the Foundation:

— giving *A Course In Miracles* first priority.
— one hour morning and evening meditations.
— a half hour during the day to step out
of the pressures of circumstances
and hold hands with God.
— one day of silence a week.

We have embarked on a full-fledged preparation
for the life of the Spirit —
a lifestyle consistent with Divine Laws.

Intellectual learning is just mere words.
A Course In Miracles is to be lived, not read.

The wise men of old
were masters of meditation,
and one with the One.

Without attracting attention
they acted out of their stillness.

One who is in the stillness
illumines himself.

Such a being is watchful,
as though it were important
not to hurt anyone.

Lao Tzu

To live is to live in innocence and to be vulnerable.
It requires a reversal of thinking
to be the extension of the Grace of God
rather than be the fear and insecurity of the earth.

Miracles are everyone's right,
but purification is necessary first.[6]

Purification begins the change of lifestyle.
It prepares one for a rebirth.

Purification.
We must cleanse ourselves of reactions,
judgments, opinions, and assumptions,
and all that originates with "ifs" and "buts" —
even the unreality of past and future in which we indulge.
And there is the need to be cleansed of prejudice.

Replace knowledge by wisdom,
and your ignorance and worry will vanish.
Do you perceive the difference
in following habit
or following conscience?. . .
The masses follow their lust and desires
having manifold wants and aims . . . .
The awakened is unwanting. . .
The masses are busy and
chase after progress.

Lao Tzu

Let us start with eliminating the unessential.
Irresponsibility makes man casual and wasteful.

We here are coming to the wisdom of conservation,
putting an end to any waste of food,
money, gasoline, electricity, water, energy,
and telephone calls.
Simplicity is the key to a new way of life.

We never buy more than we need.
We never need more than we use.
We never use more than it takes to get by
'til we learn to need less.
Chinese Proverb

We are getting our lives in order, debts paid,
and we are discovering
that simplicity is free of indulgences.
Outlets fall away when purpose and direction are clear.

PURPOSE IS MEANING.[7]

Do you realize the crucial problem
in this externalized and stimulated age is:

— lack of discipline, and
— man's relationship with food?

The awareness of this is helpful,
but awareness is different than learning.

One who is aware is the awakened one,
united with God.
Awareness is of Heaven.

Each person has to select the food that is right for him.
For the diet is determined by the state of Being,
and one's evolvement and direction.

If one is with the appointed purpose,
there is not the conflict.
What would be the diet of a person free of conflict?
Obviously the quality of his diet is different.

Discipline is not supposed to be
pre-determined and imposed.

All real pleasure comes from doing God's Will.
This is because not *doing it*
is a denial of Self.[8]

Knowledge is of things external.
Wisdom is Self-learning;
but "Know thyself" requires the space of non-judgment.
Would you find out what your mind gives its energy to?

Let us be concerned with the change
of values and right livelihood
and not just the diet.

One-To-One Relationship

At the level of the individual
is where application can occur.
That is why the workshops are over —
for they served their purpose —
and now there is the space
to be with you individually.

We offer friendship for life.
Herein lies the importance of the one-to-one relationship:
to break away from the culture of fear and insecurity.

Know, blessing is never pressured.
This one-to-one relationship
is meant for us to come to a dialogue,
to go beyond thought
to an all-knowing, impeccable State of Being where:

I AM STILL AS GOD CREATED ME.[9]

God shines His Face upon those
who hold hands and meet in HIS NAME.

And now, when everything external is questionable,
the so-called high standard of living
has become too high a cost to pay.
Most people in this affluent society are broke,
devoid of the potentials and resources
to be self-reliant.

We realize the importance of goodwill,
charity, and the Love of God.

There is the need of purified language
and absolutely clear one-to-one relationship.

> *Separated from the origin*
> *there is talk of humaneness and justice;*
> *where wisdom gives way to smartness,*
> *cunning and lying appear.*
>
> Lao Tzu

> *Separated from the All-relatedness*
> *and All-union, man seeks a substitute*
> *in human relationships and family ties.*
> *When consciousness of the unity*
> *of mankind vanished,*
> *clans and peoples and feuds*
> *without end arose.*
>
> Lao Tzu

Now there is the urgency for each to stand on his own feet
and discover the God-given holiness.

The empty talk of cosmic consciousness,
kundalini, auras, and meditation —
for sale everywhere —
has made its round and had its day.

Listen to the wisdom of our forefathers:

> *The man who never looks into a newspaper*
> *is better informed than he who reads them,*
> *inasmuch as he who knows nothing*
> *is nearer the truth than he whose mind*
> *is filled with falsehood and error.*
>
> Thomas Jefferson

Once you know:

I am sustained by the Love of God.[10]

you see not the physicality of a tree or of a human being,
but the Reality —
the totality and perfection of God.

It is being liberated from time.
Time, with its pressures, projects, and ideas,
distorts the eternity of our Reality.

Faith is not external.
It is not of the body senses.
Faith is non-verbal.
It precedes thought
for thought is reaction to something external.

Faith is without words.
It is the origin — the Truth.
It is said:

> *The sages searching for their heart with wisdom*
> *found out the bond of Being in Non-Being.*
> Rig Veda

Will you heed the call?

END YOUR EXTERNAL LEARNING.

> *We are in a society rich in means,*
> *poor in purpose.*
>
> *Society created the system,*
> *and now we have become captive of the system.*
> E. F. Schumacher

It takes a rare individual to step out
of the conceptual fantasies of it all,
and for that he needs to change his LIFESTYLE.

This is the challenge of challenges.
We settle for adjustments;
but nothing less than rebirth is required.

> *When a stupid man is doing something*
> *he is ashamed of,*
> *he always declares that it is his duty.*
> George Bernard Shaw

Behold the profile of our times:

> The crust of the earth
> and human consciousness are related.
> The thin layer that surrounds the globe —
> where life is —
> is an extension of human consciousness.
> As violence, distrust, decadence, panic,
> and the craze for armaments increase,
> life on the planet is affected.

Thus, each one of us is responsible
for his thought, deed, and action.

> *By this shall all men know*
> *that ye are my disciples,*
> *if ye have love one to another.*
> John 13:35

The Bible says:

> *And God said,*
> *Let us make man in our image,*
> *after our likeness:*

and let them have dominion
over the fish of the sea,
and over the fowl of the air. . .
and over every living thing
that moveth upon the earth . . . .

. . .Behold, I have given you
every herb bearing seed
which is upon the face of all the earth,
and every tree, in the which is the fruit
of a tree yielding seed. . .
Genesis 2:26-29

Fear and insecurity are in the blood of man now;
the mental pressures are their outcome.
Dependence is exploitable.

Probably eighty percent of the world population
is producing unessential things.
We live in a culture that abides
by the law of wishes.

The consequences of affluence and routine life
are staring us in the face.
The system fails and you fail.
You don't know what to do.
So, the plague of unemployment is upon us,
and those who have given their life energy
to the artificial and the unessential
are now self-spent,
faced with their irresponsibility.

We live by the law of wishes.
This becomes so destructive that unless we CHANGE
we become dependent on the collective consciousness.

You have to exercise your mind, your brain, your heart to find this out, because if we don't we are going to destroy each other. This is what is happening in the world — one nation becoming more important than the other, one tribe more important than the other, building up armaments — you know all that.

J. Krishnamurti

The basic issue is with yourself.
The external is abstract.
The only reality is YOU.
The action begins with you.
You are the altar of God on earth.

There is urgency to come to inner integrity.
We cannot rely upon the abstract and the external —
for it is so heavily conditioned
and bound to thought forms and belief systems.
It stamps the mark of littleness on us.

Every decision you make
stems from what you think you are,
and represents the value that you put upon yourself.
Believe the little can content you,
and by limiting yourself
you will not be satisfied.
For your function is not little,
and it is only by finding your function
and fulfilling it
that you can escape from littleness.[11]

This is why the awakened man turns
inward and rules the senses.
He lets go what passes and
values what lasts.

Lao Tzu

I AM DETERMINED TO SEE.[12]

What else is there to see
but one's own deception?
The rest is God's perfection.
The most difficult challenge
is the unwillingness to see one's own illusions,
the unrealities of one's own life.

WE HAVE TO COME TO THE REMEMBRANCE OF GOD.

Within one year *A Course In Miracles*
will change our very breath and quality of blood.
You will feel as if you do not even have a body.

This is the year for us to prepare
and relate with the eternal laws of Life.
Make space and discover your wholeness daily.

There is so much cruelty in the world
that those who are cruel do not even know it.
It is difficult for people caught in time —
and most of us are —
to not be stimulated.

So, the wise would have to discover
for himself what is harmful,
and then come to clear recognition
that time is violence.
For at the time level one is regulated by experience,
thus the more pressured we are by time,
the more violent and self-destructive
become our tendencies.

While time bound, we remain caught in illusions
and ignorant of infinity.

First, we must recognize that we are subject
to earth forces and human conditioning.
The challenge of Now is to ward off
these adopted tendencies.

Now we have the year to be free of this
and to realize the significance of urgency —
our individual responsibility and action.

How few ever rise to conviction and
integrity and stay with rightness;
how many fall and go for success
and become the destructive residents of this earth.
Whereas the appointed function
of each human being is to bring
the Kingdom of God to earth and live by:

LOVE YE ONE ANOTHER.

Behold the wholeness of Stillness.

Hopefully, it is beginning to dawn
that the action is of Grace —
and the Lord does His Work.

To outgrow the unreality of time
is to outgrow the unreality of experience
that makes one captive of physicality.

To Be is — As Created,
God Nature — Non-separated.
The Truth.
The Eternal and the Infinite.

MY SELF IS RULER OF THE UNIVERSE.[13]

As the purity of Grace begins to dawn,
one feels as if contaminated
by any projections of wishes and wantings.

God is in everything I see
because God is in my mind.[14]

Would you please give honesty to the lesson:

Above all else
I want to see things differently.

You may wonder why it is important to say, for example, "Above all else I want to see things differently." In itself it is not important at all. Yet what is by itself? And what does "in itself" mean? You see a lot of separate things about you, which really means you are not seeing at all. You either see or not. When you have seen one thing differently, you will see all things differently. The light you will see in any one of them is the same light you will see in them all.

When you say, "Above all else I want to see this table differently," you are making a commitment to withdraw your preconceived ideas about the table, and open your mind to what it is, and what it is for. You are not defining it in past terms. You are asking what it is, rather than telling it what it is. You are not binding its meaning to your tiny experience of tables, nor are you limiting its purpose to your little personal thoughts.

You will not question what you have already defined. And the purpose of these exercises is to ask questions and receive the answers. In saying, "Above all else I want to see this table differently," you are committing yourself to seeing. It is not an exclusive commitment. It is a commitment that applies to the table just as much as to anything else, neither more nor less.

You could, in fact, gain vision from just that table, if you would withdraw all your own ideas from it, and look upon it

> *with a completely open mind. It has something to show you; something beautiful and clean and of infinite value, full of happiness and hope. Hidden under all your ideas about it is its real purpose, the purpose it shares with the universe.*
>
> *In using the table as a subject for applying the idea for today, you are therefore really asking to see the purpose of the universe. You will be making this same request of each subject that you use in the practice periods. And you are making a commitment to each of them to let its purpose be revealed to you, instead of placing your own judgment upon it.*[15]

But in the world of man,
the perspective is distorted.

God created all things in perfection,
and the Child of God does all things
in reverence and remembrance of Him.
This is the definition of conservation.
Where there is conflict
there is not the gratefulness of conservation.

God is in everything I see.[16]

The bird cannot Be without the air,
the light of the sun,
nor without the tree the earth provides.
Neither can you or I Be by ourselves.
Who is not sustained by the Love of God?

NOTHING REAL CAN BE THREATENED.
NOTHING UNREAL EXISTS.

The truth of this we have to REALIZE.

The Forty Days in the Wilderness

The Forty Days was beyond the ability of words.
Every person stayed for the entire time
and was deeply affected.
For each of the participants,
it was a once-in-a-lifetime experience.

The participants undertook to give
A Course In Miracles first priority for a year,
and thus carried the blessed atmosphere with them.

Those who have taken the workshops and retreats,
to whom this *HOLDING HANDS* is sent,
too can invoke the remembrance of God daily
by giving *A Course In Miracles* the first priority.
We would be glad to send guidelines
for the daily rhythm.
You too may join us and break away
from the culture of fear and uncertainty.

The Prayer Vigil enclosed* in this *HOLDING HANDS*
is for that purpose —
so that you may set time aside for the Holding Hands
and coming to Wholeness.

The format we found most beneficial
is that of *A Course In Miracles,*
for its curriculum is of Divine Origin,
and inherent in it is the blessing to make it possible.

---

*A guideline for prayer similar to the one which appeared in *Excerpts from the Forty Days in the Wilderness* was included. See page 269.

Rejoice in the fact:

Salvation is a promise, made by God,
that you would find your way to Him at last.
It cannot but be kept.[17]

What is understanding?
Understanding ends the activity of thought
and is a moment of freedom,
the Holy Instant that perceives
the perfection of creation.

If you do not have *A Course In Miracles,*
acquire a set of the three books.*
It will correct misperceptions.

This is a glorious year.
Join us in our undertaking.

The power of decision is my own.[18]

There are the difficult times ahead.
This is obvious.
The purpose of being aware of this
is the exact opposite of being panicky
and feeling helpless,
because man is never helpless.

We have to know what the internal and
external world is in order to master it.
The only purpose of experience is to outgrow it.

For it is the self-actualized human being
who will break out of collective culture.

**A Course In Miracles* is now available in a combined softcover edition as well as in the original, three volume, hardbound edition. (Editor)

It is when we put our reliance on the externals
that we are dependent — bound to inadequacies.

Confidence is something born out of
one's own integrity and character.
These are attributes of the Spirit.

This *HOLDING HANDS*
is a call and a cry in the wilderness.

MAKE STRAIGHT THE WAY OF THE LORD.
John 1:23

The Forty Days has been the greatest
accomplishment of this life.
Now that we have made the contact,
and we have established the bonds,
I intend to spend a year in contemplation
for the new to BE.

The place has been found.
By the time this reaches you,
I will be with the new rhythm.
It is always possible to find
the space and the resources to change.

Reading this,
you too become part of my retreat.

You, who have taken the workshops and retreats,
can write. If it is urgent,
I will heed and respond to your call.

When you call upon me,
this is the first thing I will ask:
"What is your lesson of *A Course In Miracles* today?
For inherent in the lesson
is the answer to your problem."

I see the importance of not overextending
and getting too organized,
but to be available for those who are serious
for the one-to-one dialogue.

It is the non-commercialization of friendship,
where the two meet in His Presence
to bring His Will to earth

. . .*as it is in heaven.*

This is the appointed function.
Insecurity does not touch it.
It is of love.

We value the one-to-one relationship with you —
the friendship of holding hands.
It is not one way.
We welcome your input
and would like to hear from you.

From the Foundation for Life Action,
we now rise to the individual, one-to-one
relationship of *Love ye one another.*

Where there is love,
the Name of God cannot be commercialized.

Now in our life,
let us turn to the invocation of prayers.
For prayer is man's real strength.

Let us invoke daily in our prayers
the law of Love.
Pray for peace and goodwill
towards all men of all nations.
Thus surround the world with your blessings.

But make not your prayer a ritual.
Step into an impeccable space within yourself,
and then PRAY unhurriedly and lovingly.

It is the stillness that brings about
the shift of energy and becomes the prayer.

Prayer is the voice of the Holy Instant
of one's wholeness
that imparts the blessing.

Make:

By grace I live.
By grace I am released.[19]

your Reality.

Prayer is the medium of miracles.
It is a means of communication
of the created with the Creator.
Through prayer love is received,
and through miracles love is expressed.[20]

Let us pray for each other.

*Summary**

Eminent and renowned men talked about
non-violence and de-centralization.
They warned us about the exploitation
and the catastrophe of the concentration of power
in the hands of few.

They pointed out that
with the increase of industrialization
and the commercialized communication media,
the nations of the world would turn
from being a means for the welfare of their people,
to an end in themselves.

They voiced the consequences of artificial life
and the irresponsibility of waste.
Their appeal was to the masses,
for they had a cause.
But now we see that "the masses" is an abstraction
born of their own partial reality.
Because they sought leadership,
they created "the masses."

The so-called great economists,
statesmen, reformers, and scientists
have not succeeded in harmonizing and humanizing society.
Their "society" is as abstract as "the masses."
Besides, at what price is their help
with their misconceptions of a future

---

*For a further exploration of the topics presented in this section, see *The Future Of Mankind — The Branching Of The Road* by Tara Singh, published by the Foundation for Life Action in 1986.

which is only the extension
of the unreality of the past?

The question is:
Can humanity be helped externally?

These men of intellectual expansion,
in themselves, are not whole
but are caught in the surface phenomena
of appearance and causes.

Their observations are at the level of duality.
They promote change from "this to that" —
the opposite.
But is that a change at all?
Is it not just modification?

The world ideas are influences and not energetic
even though they may sway people
with their goals and projects.
They do not have the purity
of direct dialogue with an individual
where the Divine is present.

Dialogue where two or more
gather in His Name, is different.
It dissolves words.

Dialogue is religious
where the intensity is to realize
the Truth that ends the words.

Men who are whole and not caught in the darkness
of their own unrealization —
God-lit — Buddha, Jesus, Mr. J. Krishnamurti.
They dealt with the individual,
for they were of the Kingdom of God.

Men in blossom, having the love within them,
have the space for the one-to-one relationship.
For they are the Masters of Self, living in peace.

They do not commercialize for they are of detachment.
You must have the discrimination to discover them,
and an uncompromising passion for the Truth,
and no vested interest.
For, the so-called learning can be a preoccupation.

Will we now listen to those
who are free of belief systems,
who are timeless,
and see man as an extension of eternity
made in the Image of God
and not as a citizen?

Their peace is of infinite mind
that transcends all that is external.
They are of the indivisible Law of the One
for they see but the One Reality that God IS.

The God-lit man through whom
the Heavens speak directly to the earth —
it is the largeness of his reverence
that relates us to non-duality.

For example,
A Course In Miracles is not a translation
and is free of interpretation.
It is the order of the first step first —
the systematic Divine order of education.
Inherent in it is the blessing
to bring you to the truth of it —
the Divine Laws that impart their strength.

Then there is Mr. J. Krishnamurti —
the first world teacher.
His silent mind goes to words
and inspires us to stillness.
And if we have the ears to hear,
then we too journey with him
to our own Divine silence.
He does not distract or create dependence
but relates us with our own perfection.

Today, daily, millions of dollars worth
of free publicity is given
to violence and disharmony among nations
and the omen is of sirens wailing in the streets
of a nation that was meant to honor:

"IN GOD WE TRUST."

Where is the space for the trust
and the stillness of wisdom?
For who can renew external life
without virtue and peace within?

We have infested the world with separation
that prevents the individual from awakening
to his own Divine Self,
that can usher in the new consciousness of:

LOVE YE ONE ANOTHER.

A person needs to be a strength to another.
But this is achieved by the mastery of life within.

We have the resources and the capacity to change.
No one need be pressured into helplessness.
Change is an internal action
that starts with self.

Man is of Divine Origin.
When we undermine that,
we undervalue the action of Life Itself.

When I am transformed,
the world is affected.

I am
the whole,
the totality of the universe.

Part VI

Non-Commercialized Life
The Path of Virtue

Introduction

It is important that you read this *Holding Hands*
in the quiet holiness of your own space.
Let us learn to sit still and upright
to receive the light of truth within.

There is a state that touches the words
but is not of words.
It is the quality of reverence
we give to the moment
that invokes the blissful safety
of the Mind of God of which we are a part.

The year retreat that we offer
IS NOT MY PLAN.
IT IS NOT MY PROJECT.
IT IS:

God's plan for salvation.[1]

The light is in it.
Thus, it is not commercialized.

The gift of internal awakening
surpasses all blessings.
Inherent in it is the benediction
that removes the obstacles
and the limited double vision of doubt.

Welcome it with joy!
It will awaken your potentials

and make the way clear,
for it deals with your life purpose.
Thus it cannot be blocked by external circumstances.

To give and to receive are one in truth.[2]

Give yourself the space
to realize the truth of it.

Truth will correct all errors in my mind.[3]

You are responsible for your confidence.
Confidence has the power to awaken you,
and to let God work in and through you.
The means are provided
if you do not project or conclude.

Truth silences the mind,
stops time
and imparts tremendous energy.
Be with it.

There is nothing my holiness cannot do
because the power of God lies in it.[4]

I am in the spirit of Gratefulness.
My mind extends its joy and strength to you,
so that you can be part of:

God's plan for salvation.

Know the "PLAN" itself has the energy
to bring people together.

Non-Commercialized Life
The Path of Virtue

For you, who *will* to change, we offer:
the One Year, full-time study of *A Course In Miracles.*
It is for the serious minded
who are dedicated to being a part of:

God's plan for salvation.

For a year, from Thanksgiving, November 25, 1982,*
a residential study program
of *A Course In Miracles* will be offered.
There are no tuition fees,
for the Name of God cannot be commercialized,
and we gather for a real purpose.

Behold, in this day and age,
how rare the person who has the space
and the resources to be part of
a life-changing experience.
In prior societies,
this was seldom the question.
People walked out of kingdoms to be free.

Liberation was important,
and man's relationship with God was important.
Today, man is bound hand and foot.

---

*The One Year Non-Commercialized Retreat: A Serious Study of *A Course In Miracles* was delayed. It took place in Los Angeles, California, from Easter Sunday, April 3, 1983, to Easter Sunday, April 22, 1984. (Editor)

However, for those who are earnest
and have order in their lives,
it is still possible to respond
to that which is righteous and timeless.

THE WILL OF MAN HAS ITS OWN SPACE.

TRUTH IS NOT RULED BY TIME.

The discovery of light and strength
in the words of the Course is important.
Give your mind and heart to it.

I am as God created me.[5]

and

I am under no laws but God's.[6]

A Course In Miracles,
with its three hundred sixty-five lessons,
is for a year
and in an atmosphere
disentangled from the world,
we will give it our full attention.

Wherever the Thought of God is shared,
the atmosphere is holy.
Where two or more meet in His Name,
they are blessed.

In my one year of silent retreat
it has been witnessed:
If your attention is with it,
the daily lesson of *A Course In Miracles*
will unfold all day long.

And at night, too,
its grace is upon you while you sleep.
Inwardly you are intimated about the rhythm
of the exercise of the day.
It is completed when you awake.

Within twenty-four hours,
the truth of the lesson can be recognized.
Eagerly, the gladness in you reaches
for the next lesson.
It becomes spacious, an internal action
of gratefulness
and not an intellectual practice.

I thank my Father for His gifts to me.[7]

The daily lesson of *A Course In Miracles*
is to be practiced
in the spirit of gratefulness.
You will see.
TIME STOPS,
even during the moments of practice
between the hours.

It has nothing to do with how busy you become. . .
It is the intent and the joy of reminder
that breaks the barrier,
and brings you in contact
with your eternal identity.

Each lesson is a Holding Hands with God —
an eternal moment that extends itself.
Give it the space of your own gladness
and feel the impact in every cell
from head to toe.
Be grateful to the Course —
that IT is of help and service.

A Course In Miracles, I was intimated:

". . .is not a matter of words.
The Course has to be lived,
or it will pass you by like a breeze.

You have to want something very, very much —
and not just like to talk about things.
The Course is for people
interested in responsibility
and who think things through.
Christ's Thought is not of levels.
It sees only the One, the Truth.
God is in all that IS.

You have to be on the right track
for Him to answer you."

Training of Teachers

The one year is for the training of
teachers for the ministry of God.
TEACHERS WHO LIVE:

. . .under no laws but God's.[8]

It is a very auspicious and powerful Action,
for it focuses the planetary forces
on the earth.

Only God's plan for salvation will work.[9]

I need devoted teachers
who share my aim of healing the mind.[10]

His Assurances

Remember this:

These things I have spoken unto you,
that in me ye might have peace.
In the world ye shall have tribulation:
but be of good cheer;
I have overcome the world.
John 16:33

If you are willing to renounce
the role of guardian
of your thought system
and open it to me,
I will correct it very gently
and lead you back to God.[11]

I will teach with you
and live with you
if you will think with me,
but my goal will always be
to absolve you finally
from the need for a teacher.[12]

You are not at peace
because you are not
fulfilling your function.
God gave you
a very lofty function
that you are not meeting.[13]

I will come in answer
to a single unequivocal call.[14]

Let us ask the Father
in my name
to keep you mindful
of His Love for you
and yours for Him.
He has never failed
to answer the request,
because it asks only
for what He has already willed.[15]

I have said already that
I can reach up and bring
the Holy Spirit down to you,
but I can bring Him to you
only at your own invitation.[16]

If it helps you,
think of me holding your hand
and leading you.
And I assure you
this will be no idle fantasy.[17]

It is possible, in a year, to step out
and come to a state where:

Nothing real can be threatened.
Nothing unreal exists.

What is God's Plan for Salvation?

It is an Action of God
where time passes into Eternity —
wholly true and unlimited.

God's Plan for Salvation is without opposite.
It is an extension of God's Love
that ends the insanity of separation.

God's Plan for Salvation is God's Plan.
It is not yours; you are part of *it,*
for God and His Son are not different.

God's Plan for Salvation has begun
and is already in effect.

It works through the human being.
It is the human being that desecrates,
and it is the human being
that brings the Will of God to earth.

Let us make a place ready
and remove all the meaningless
and self-made substitutes
which we have placed on the holy Altar of God
within our mind.

Thought must inevitably prevent action.
This you need to realize
to be a part of God's Plan.

God's Plan for Salvation
is not subject to anything external.
It introduces man to his own Eternity
rather than responding to events on earth.

God's Plan for Salvation
is a continuous Newness
without past or future.
Thought cannot touch it.
You have to become part of it
to understand it.

Its renewal brings man
to the perfection of the present
in which there is no unfulfillment because:

The Kingdom is perfectly united
and perfectly protected,
and the ego will not prevail against it.
Amen.[18]

We have to understand
God's Plan for Salvation.
In the very understanding of it
we are liberated.

You do not have to do anything.
If you could only come to the clarity
of what IT is,
IT will broaden your perspective
and transform your life.

GOD'S PLAN FOR SALVATION.
The blessing of all God-Incarnations —
for They are One —
is behind it —
the Dawn of the New Age.

It is possible that anxiety, fear, and blame
can be dissolved in the first one hundred days.
This is significant,
for unless insecurity is removed,
it may not be possible to undertake anything
that is not self-centered.

From the Gratefulness Journal
of my Retreat:

> "You arrive at the hundredth lesson
> and blossom with joy
> as your will unites with the Will of God.
>
> Confusion over,
> you behold the support of the Universe
> behind you and observe
> how each day's lesson
> kept unfolding within you,
> in spite of yourself,
> because you let it be.
>
> GOD WILLS THE COURSE TO SUCCEED."
>
> > *. . .you cannot fail in your efforts to achieve the goal of the course. You will see because it is the Will of God. It is His strength, not your own, that gives you power. It is His gift, rather than your own, that offers vision to you.*[19]

It is a reality.
And you abound in your gladness,
eager to share.

My part is essential
to God's plan for salvation.[20]

"By the hundredth lesson,
you can glimpse the fulfillment of your life,
the joy of creative productivity
rather than meaningless existence.

Anyone who takes *A Course In Miracles* seriously
will have the love to share,
laughter to laugh,
the strength of certainty
and a productive life,
untouched by insecurity."

We are not talking about the preoccupation
of teaching and external learning,
but an awakening.
Awakening expands the space between the thought.
Miracles are not ruled by time.

Truth will correct all
errors in my mind.[21]

My mind is part of God's.
I am very holy.[22]

The one year offers the path of virtue
of non-commercialized life
in exchange for mundane existence.

The vitality of new action is upon us.
Nothing is difficult that is wholly desired.
You need space to be yourself.

Let me recognize my problems
have been solved.[23]

Today the time of light begins for you and everyone.
It is a new era, in which a new world is born.[24]

"How blessed the man who has no desire.
Desire is always personal.
It is the acknowledgment of separation."

I am safe today
because there is no will but God's.[25]

I am under no laws but God's.[26]

And His are the laws of freedom.[27]

I am the means God has appointed
for the salvation of the world.[28]

To those Jesus sent forth, He commanded:

. . .as ye go, preach, saying,
the kingdom of heaven is at hand.
Heal the sick, cleanse the lepers,
raise the dead. . .
freely ye have received, freely give.
Provide neither gold, nor silver,
nor brass in your purses,
nor scrip for your journey. . .

And into whatsoever city or town
ye shall enter,
inquire who in it is worthy. . .
And when ye come into an house,
salute it.
And if the house be worthy,
let your peace be upon it:
but if it be not worthy,
let your peace return to you.
And whosoever shall not receive you,
nor hear your words,

when you depart out of that house or city,
shake off the dust of your feet. . .

. . .take no thought how or what ye shall speak:
for it shall be given you in that same hour. . .
For it is not ye that speak,
but the Spirit of your Father
which speaketh in you.
Matthew 10:7-20

My voice is His, to give what I receive.[29]

Salvation is my only function here.[30]

My part is essential
to God's plan for salvation.[31]

From my Gratefulness Journal:

"Meditation is an atmosphere of God.
I sit quietly by my own Holiness.
The present extends into timelessness.

This surely is the plan
and the direction for man.
No guidance is necessary as verification.
What is required is the dispelling
of our own hesitation,
the alternative which causes dependence
and uncertainty.

What reason is there to hold my doubt as truth?
Doubt demanding clarity is the block to IT.
For clarity already IS the benediction —
a natural state of being.
Application is instantaneous.

Desires, choices, and alternatives
emerge out of dullness and reluctance.

There are no questions or answers.
The certain are ever calm.

The Holy Spirit is the working companion.
He will accomplish whatever is needed
for:

God's plan for salvation.

I must move from the totality of trust
and experience actuality itself unfold.
Be with AS IS,
absent only from the future and the past,
free of belief."

. . .*you* are *responsible for what you believe.*[32]

Belief is never real.
But man is caught in it.

Remember this:

I am here only to be truly helpful.
I am here to represent Him Who sent me.
I do not have to worry
about what to say or what to do,
because He Who sent me will direct me.
I am content to be wherever He wishes,
knowing He goes there with me.
I will be healed
as I let Him teach me to heal.[33]

"This State is a space that encompasses
the world and surrounds the universe.

Truth too is a state that is constantly
bringing space into this time bound planet.
Truth is not ruled by time.
The Man of God is not ruled by time
because he is spacious.

Yes, there is an energetic
STATE of Holiness,
that takes place in the Action of Love,
independent of the movement of thought.

It is the dawn of authentic words,
the Thoughts of God,
eternal in their consistency —
the creative state of who you are."

Light and joy and peace
abide in me.[34]

To everyone I offer quietness.
To everyone I offer peace of mind.
To everyone I offer gentleness.[35]

There are three classes of men:

— Those who are the extension of God —
God-lit men who live by Eternal Laws
whose peace fills the universe.
They do not commercialize life.

— Those who organize.
The secondary men of vested interests.
Men of causes.

— Those who interpret.
The belief-ridden multitude
that rally around the organizers.

The Extensions are men who extend Holiness
and do not deviate into the external.
Neither Lord Buddha,
nor the Christ charged fees,
nor in this day and age did Mr. J. Krishnamurti,
whose words are eternal.
He did not even accept royalties from his books.

The author of *A Course In Miracles* remains anonymous,
as the Light of Truth behind appearance.
She remains untouched by words and
bestows upon mankind Thoughts of God —
Truth that is not ruled by time.

And here we have the atmosphere of the
Thoughts of God
offered to us for a year free of charge.
Will you heed the call?
Put away your external learning,
and come not to *learn* but to *BE.*

The Teacher at the Retreat is the PRESENCE.

I am not the teacher,
but a messenger of gratefulness,
a gift of my Father to us all.

The One Year Retreat is a unique event.
It is not for the camp followers.
You have to merit it.
We will be selective.

It is for the austere —
those committed to the impersonal life of the Spirit,
a mind capable of renewing itself.

Assurances

I am one Self,
united with my Creator.[36]

You are one Self, united and secure in light and joy and peace. You are God's Son, one Self, with one Creator and one goal; to bring awareness of this oneness to all minds, that true creation may extend the Allness and the Unity of God. You are one Self, complete and healed and whole, with power to lift the veil of darkness from the world, and let the light in you come through to teach the world the truth about yourself.[37]

. . .your mind has found the function that it sought to lose. Your Self will welcome it and give it peace. Restored in strength, it will again flow from spirit to the spirit in all things created by the Spirit as Itself. Your mind will bless all things. Confusion done, you are restored, for you have found your Self.[38]

I am the Son of God. No body can
Contain my spirit, nor impose on me
A limitation God created not.[39]

I will accept my part in God's plan for salvation.[40]

What can my function be but to accept
The Word of God, Who has created me
For what I am and will forever be?[41]

Begin by asking Him Who goes with you upon this undertaking that He be in your awareness as you go with Him. . . . You speak to Him today, and make your pledge to let His function be fulfilled through you.[42]

THE JOURNEY BACK
The Direction of the Curriculum
[For the One Year Retreat]

Knowledge is not the motivation for learning this course. Peace is. This is the prerequisite for knowledge only because those who are in conflict are not peaceful, and peace is the condition of knowledge because it is the condition of the Kingdom. Knowledge can be restored only when you meet its conditions. This is not a bargain made by God, Who makes no bargains. It is merely the result of your misuse of His laws on behalf of an imaginary will that is not His. Knowledge is *His Will. If you are opposing His Will, how can you have knowledge?*[43]

There is only one curriculum
consistent with God's Will in the world,
and it starts with the undoing —
the journey back to one's own perfection.

This is a course in miracles.
It is a required course.
Only the time you take it is voluntary.
Free will does not mean that
you can establish the curriculum.
It means only that you can elect
what you want to take at a given time.[44]

The goal of the curriculum . . . is "Know thyself." There is nothing else to seek. Everyone is looking for himself and for the power and glory he thinks he has lost.[45]

God's Will is perfect happiness for me.
And I can suffer but from the belief
There is another will apart from His.[46]

I have a function God would have me fill.[47]

Let me remember I am one with God,
At one with all my brothers and my Self,
In everlasting holiness and peace.[48]

In quiet I receive God's Word today.

Let this day be a day of stillness
and of quiet listening.
Your Father wills you hear His Word today.
And so He calls
from deep within your mind
where He abides.
Hear Him today.
No peace is possible until His Word
is heard around the world;
until your mind, in quiet listening,
accepts the message that the world must hear
to usher in the quiet time of peace.[49]

To fulfill the Will of God perfectly is the only joy and peace that can be fully known, because it is the only function that can be fully experienced.[50]

"IN GOD WE TRUST"

Part VII

Invitation to the One Year Non-Commercialized Retreat

Invitation to the One Year Non-Commercialized Retreat

My single purpose offers it to me.[1]

Times will reveal themselves —
that you cannot depend on the externals
and without the externals
there is no personality.

The time is coming when the fear of the externals
will destroy the external.
It is a very painful thing
because we identify with unreality.

There is no relationship at the personality level.
It is only the illusion of living
and not the extension of Life,
the holiness of eternity.

Anything you and I, as idiosyncrasies, would do
would emerge out of expedience or deception
and will inevitably be false.
Self-interest cannot afford RIGHTNESS.

Wisdom is a strength most people do not have.
Man is exhausted and dissipated of morality.

The system has drained it out of each
and now, only into the system
can we fit for our survival.
But is not the system everywhere
degenerating at an alarming rate?

In the commercialized society,
behold the abuse of man.

What most people relate to and are limited by
are their desires and wants.
In the absence of integration,
the unessentials assume enormous importance
and the human being is victimized.
Thus, all one is left with is the life of consequences.

Yet we are responsible for the purity of peace within.

Tomorrow is the Age of Panic.

> *You cannot distort reality and know what it is. And if you do distort reality you will experience anxiety, depression and ultimately panic, because you are trying to make yourself unreal. When you feel these things, do not try to look beyond yourself for truth, for truth can only be within you.*[2]

Let me remember I am one with God.[3]

For a year beginning Thanksgiving, November 25, 1982,*
a residential study program
of *A Course In Miracles* will be offered.
There will be no tuition fees,
for the Name of God cannot be commercialized.

We offer the year of serious study
to come into relationship with what is REAL.
That is why we are providing an atmosphere for a year
free of charge, pressure and interference.

That the year can occur non-commercialized

---

*The One Year Retreat actually took place from Easter Sunday, April 3, 1983, to Easter Sunday, April 20, 1984.

in a materialistic, mechanical culture,
is a statement of the eternal on the time bound.
We are doing it in order to prove it can be done,
that life need not be commercialized.

If you recognized this, you would be grateful,
and nothing could keep you away,
for it imparts the energy and grace to bring you to it.
But if you are blocked by involvements and problems,
then you have not recognized the truth.

It is a statement that is universal,
an Eternal Law that affects the time level.
Its vitality is its reality.

It is an experiment of Newness —
something that has not been done before.

The Name of God
Cannot Be Commercialized.

MY SINGLE PURPOSE OFFERS IT TO ME.

Single Purpose cannot be with anything that is unessential.
If we are with the Single Purpose,
then each one of us becomes creative and responsible.
Each person has a function according to Divine Plan.

You do not recognize the enormous waste of energy
you expend in denying truth.[4]

Personality issues prevent us from being part of:

God's plan for salvation.

But to a CHANGED life, everything is accessible.

The recognition of *I am as God created me*[5]
eliminates the helplessness.

> *All things are given you. God's trust in you is limitless. He knows His Son. He gives without exception, holding nothing back that can contribute to your happiness. And yet, unless your will is one with His, His gifts are not received. But what would make you think there is another will than His?*[6]

> *It is a new and holy day today, for we receive what has been given us. Our faith lies in the Giver, not our own acceptance. We acknowledge our mistakes, but He to Whom all error is unknown is yet the One Who answers our mistakes by giving us the means to lay them down, and rise to Him in gratitude and love.*

> *And He descends to meet us, as we come to Him. For what He has prepared for us He gives and we receive. Such is His Will, because He loves His Son. To Him we pray today, returning but the word He gave to us through His Own Voice, His Word, His Love:*

> *Your grace is given me. I claim it now.*
> *Father, I come to You. And You will come*
> *To me who ask. I am the Son you love.*[7]

This is what *A Course In Miracles* says,
and it is not disputable.
Have you tried and come to application?
Is it not worth a year of your life to discover
the truth of it — an awakened state?

Now to the Mind of the Age,
having lost the meaning of words,
is given *A Course In Miracles* —
the Thoughts of God.

Consistent with this,
we provide the space for newness to BE,
and the atmosphere of one year
to those who have the conviction
and are serious-minded.

A Course In Miracles brings one to unlearning.
Even though the curriculum of the course is Divine,
and you are inspired,
even then most people can hardly listen.

Your grace is given me.[8]

After hearing this what is one going to do?
Do our daily routine of the lesson?
Or be awakened to our own, full potentials
that cannot fit the routine?

Could you hear this?

> *The Call for God is heard and answered. Now has fear made way for love, as God Himself replaces cruelty.*[9]

Behold, *A Course In Miracles* gives the means
by which His Will is recognized.
His Grace is yours by your acknowledgment.

We have learned to read,
but we have not learned to recognize the truth.
So the year then is provided
to go beyond the reading to the ACTUAL STATE.

Each of *A Course In Miracles'* lessons carries with it
its own distinct vibration, and within the Year,
those who have been consistent will never be touched
by insecurity or contaminated by fear.

The Course tells us:

> *This course offers a very direct and a very simple learning situation, and provides the Guide Who tells you what to do.*[10]

Everyone would welcome then
that the space is provided
to know the truth of the lesson,
that light within the mind
that is not touched by words.
Merely reading about the state
is not the state.

THE YEAR IS FOR THOSE WHO ARE SERIOUS
AND WHO WISH TO RISE
TO THAT STATE OF THE COURSE
AND MEET IT.

I rule my mind,
which I alone must rule.[11]

I do not recognize my inadequacies as Reality.
I am determined to come to the sanity of the Law of One.

I am among the ministers of God.[12]

Let us consider what God's Plan for Salvation is,
and be amongst the Ministers of God.

The Year is for the training of teachers —
a non-commercialized life —
the Ministry of God.
It is not for those
who are going to go back to the old routine,
but for those who aspire
to be part of GOD'S PLAN FOR SALVATION.

GOD'S PLAN FOR SALVATION
means you are in HIS Service.

To struggle to survive within the framework of separation
is to be subject to the deteriorating forces of time.

Non-commercialized life is a step ahead of the external,
is not affected by inflation,
nor unemployment, nor economy changes —
because Rightness brings one to relationship
with that which is Timeless.

But we know so little about Eternal Laws,
THE PATH OF VIRTUE,
or the resources of a non-commercialized life.

We live in an age when virtue and ethics
are forgotten and lost.
And now we have our insecurities, isolations,
and fear of consequences for having lost IT.

BUT AWARENESS IS A PLACE IN YOU
THAT IS NOT OF THE EARTH.
RIGHTNESS IS THE LAW BEFORE WHICH
EVERYTHING EXTERNAL BOWS.

With Rightness you will
never be poor or lonely in your life.
In non-commercialized life
lies the salvation from the external.

The peace of God is shining in me now.

Sit quietly and close your eyes.
The light within you is sufficient.
It alone has power to give the gift of sight to you.

Exclude the outer world,
and let your thoughts fly to the peace within.
They know the way.
For honest thoughts,
untainted by the dream of worldly things outside yourself,
become the holy messengers of God Himself.[13]

The Law is:
All we have to do, and must do,
is to bring the old to an end.
That is the only responsibility.
The NEW already is.

The energy of freedom within one
awakens the mind to its own holiness.
The function God has given us
makes us the co-creator.

The Year gives us the time
and the individual holding of hands
in the atmosphere that invokes:

If it helps you
think of me holding your hand
and leading you.
And I assure you
this will be no idle fantasy.[14]

This time we are ready to give more effort and more time to what we undertake. We recognize we are preparing for another phase of understanding. We would take this step completely, that we may go on again more certain, more sincere, with faith upheld more surely.[15]

It is consistent
with the spirit of *A Course In Miracles:*

Invitation to the One Year Non-Commercialized Retreat

. . .this course was sent to open up the path of light to us,
and teach us, step by step,
how to return to the eternal Self
we thought we lost.[16]

You will always have time for other things,
but make space for the One Year.

The One Year Retreat is about the Wisdom of Life.
Organizations which are religious or commercial
want your energy and your money,
but in the Year's Retreat,
the "Unaccessible" is the Given
which you have to merit.
It is independent.

In the absence of trust,
there are only compromises.
The absence of love has its consequences.
In trust compromises end.
In love consequences end.
When the truth becomes clear,
you discover there is no other way than RIGHTNESS.
Rightness is whole and all-encompassing.
In trust and gratefulness lies the freedom of man.

The Wisdom of Life is the Reverence for Life.
In reverence there is no status, no personality,
only the perfection and the holiness of your being.

THE ONE YEAR IS FOR THE PATH OF VIRTUE.

The Blessing

Can you imagine being part of *God's plan for salvation*
knowing the certainty of your function,
and being an extension of Eternity?

The Second Coming is the awareness of reality,
not its return.[17]

We are so thrilled that a sense of blessing is upon us,
and we want to share it.
Please do not mistake our enthusiasm for persuasion.

No matter what one's situation is,
it only needs the pure intent.
The Year is possible for you unless you deny it.

In God's world there are no problems.
Man's belief does not determine Eternal Laws.
The absence of rightness breeds consequences
that make things difficult.

But the Heart of God reaches out,
assuring us it is possible.

If you feel like talking to us,
we care and would love to help explore
and make the Year possible for you.

It is the offering of friendship.

Heaven is the decision I must make.
I make it now, and will not change my mind,
Because it is the only thing I want.[18]

Bringing us to that state
is the purpose of the Year.

That is the decision,
the only decision.

The Function of the Ministers of God

Your hand becomes the giver of Christ's touch;
your change of mind becomes the proof
that who accepts God's gifts can never suffer anything.
You are entrusted with the world's release from pain.

Such is your mission now.
For God entrusts the giving of His gifts
to all who have received them.
He has shared His joy with you.
And now you go to share it with the world.[19]

I am entrusted with the gifts of God.[20]

What is Man's Function on Earth?

There is only one function for all of us —
to come to Wholeness,

TO END THE SEPARATION.

Everything in creation provides for this.

Thus insecurity, fear, and anxiety are not in place.
These only come into being
when you are not with your function,
and you can afford to project alternatives to:

I am as God created me.[21]

The separation ended,
then the NEW, the rebirth —
the extension of co-creatorship —
begins.

The ministry of God means
man, like God, walking on the earth,
changing the vibrations of the planet,
bringing peace to everything that breathes and lives.

I am as God created me.

One-to-One Year Retreat

The one-to-one relationship
in this externalized age is essential.
In this my love for you becomes my strength
and sees who you are and not what you do externally.

Thus it is not the love
that gets involved in personality.
For love cannot be reduced
to the idiosyncracies of you and me.

Would you not like the Truth to set you free?

Love expresses the sacredness of Life.
It is an internal discovery
that few men have made.

It is not the wanting of love externally.
For it already had and IS.
Such a being shares what reality IS
and is a compassionate friend of his fellowman.

He is aware but not reactive
to the inadequacies of another.
For his is the Path of Virtue
not limited to personality.

Do you see the need of this timeless atmosphere
to deal with the deepest issues of man,
to bring him to the Path of Virtue,
to which is given:

I am entrusted with the gifts of God.[22]

Registration for the One Year Retreat

It is not commercialized. There is no tuition. We wish to make you aware from the very outset that non-commercialized life is possible, for it stands on the solid ground of virtue and goodness.

Location: We are doing the job of getting you involved. Your participation in finding the place is appreciated. It is a cooperative action. Your input and your energy are needed and welcomed.

Cost: Boarding and lodging costs will be determined by where we settle and what the arrangements are. A minimum fee will be charged for Foundation services.

We need to know if you are coming. There are no deposits. We want words that are dependable. One favor that you can do is be responsible for what you say. Please write a letter. We want to know your background. Let us know what brings you to the decision to attend. Is it a wish or an outgrowing? Unless it is born out of your own incentive and has the honesty, you will not be able to see it through. It is a serious undertaking. Upon receiving your letter we will send a detailed application form.

For us the location of the Year is secondary.

The PLACE is Rightness in your life.
It is not external.

During the One Year there would be issues
one has to confront,
and this is not possible without
the love of wisdom and the strength of integrity.

The Year is not an abstract learning situation.
It is an awakening.

There will be the space for the one-to-one dialogue
so that the lesson of the day
becomes applicable at the individual level.

We have seen that the intellectual knowledge of man
is burdensome because it is limited to the senses.

We will be very selective.
It demands application.
On this we will insist.
The retreat is for those whose minds
have the vitality to renew themselves.

You come to be amongst the Ministers of God
and have chosen that to be your function in life.

This is what is meant by:

My single purpose offers it to me.

I am among the ministers of God.[23]

My part is essential to God's plan for salvation.[24]

Salvation of the world depends on me.

Here is the statement that will one day take all arrogance away from every mind. Here is the thought of true humility, which holds no function as your own but that which has been given you. It offers your acceptance of a part assigned to you, without insisting on another role. It does not judge your proper role. It but acknowledges the Will of God is done on earth as well as Heaven. It unites all wills on earth in Heaven's plan to save the world, restoring it to Heaven's peace.

Let us not fight our function. We did not establish it. It is not our idea. The means are given us by which it will be perfectly accomplished. All that we are asked to do is to accept our part in genuine humility, and not deny with self-deceiving arrogance that we are worthy. What is given us to do, we have the strength to do. Our minds are suited perfectly to take the part assigned to us by One Who knows us well.[25]

FOR WHAT YOU ARE GRATEFUL,
YOU WILL NEVER BE DENIED.

Part VIII

"I Am in Need of Nothing but the Truth"

"I Am in Need of Nothing but the Truth"

It is the encouragement of your response
to the "Invitation to the One Year Retreat"
that brings this *Holding Hands* into being.

"Dear Brother,

I was given a gift today by the Holy Spirit, an answer. An Invitation to the One Year Retreat found its way to me with divine timing. I have been working with *A Course In Miracles* for two years, and I know that it is my way home. Please send me an application for the One Year Retreat. You have my assurance that I will come."

"What makes me want to attend the retreat? I know that God is the only answer, and *A Course In Miracles* shows me the best way for me at the moment to get there."

"The One Year Retreat seems to be an answer to a prayer. The opportunity to live a life of real spiritual values among people of like mind for a year will be the beginning of a new way of life. I welcome the opportunity to share this experience."

"I long for union with my brothers with no one left out of the bath of God's Love. Reading your Invitation to the One Year Retreat touched that sense of the possible, and I cried in gratitude."

"My lesson for today is: *I give my life to God to guide today.*[1] I will be there."

"I want to attend the Year, because it is time to get to the truth."

"I want to first thank you for sending me the Invitation to the One Year Retreat. It is like nothing I have ever seen before. There are no adjectives I can apply to it — but I can only express my gratefulness to you for being who you are and that you share it with me."

"I suddenly understood the beauty of what you are doing by experiencing the year yourself before extending it to others. It reminded me of so many things: (order in my life for instance — you should see me cleaning out things!), rightness (more sensitivity and awareness). I am also aware of what a gap there is in 1) hearing once more and being reminded, and 2) application, and 3) the step that seems so *unattainable* of being in the state where one does not deviate; where rightness just *is*. . . You have given me glimpses, and I am grateful."

"Thank you very much for the information regarding the study course in the Course. It is a wonderful idea. Having lived, taught, and studied the Course for five years, you can imagine how supportive I am of your work. . . I just felt inspired to write to you, to send you my love and light and laughter and support for your program."

"The offering to attend the One Year Retreat is seen to me as a direct experience of how the universe is taking care of me."

"I was so glad to hear of the news of the Year Retreat. Be advised of my intent to be at the retreat. There could be no greater joy."

We could go on with these shared feelings.

"I Am in Need of Nothing but the Truth"

We are enthralled by those individuals
who see the value of giving a year to God.

It confirms that,

God's plan for salvation

is God's Plan.
It is not self-projected.
"It" belongs to no one
and is shared.

Please LISTEN.

This non-commercialized retreat is not
a personality plan or pretense.
It has the authority of Life and
is gathering strong momentum.

In the midst of chaos and changes,
its compassionate austerity stands
unmoved and independent.
The false disappears before its sacredness,
for it has its own vibration.

It calls upon your pure and austere state
which has never been soiled by
the dreams of personality.

The Name of God
Cannot Be Commercialized.

My single purpose offers it to me.[2]

My part is essential to God's plan for salvation.[3]

Rightness

We have to learn
Rightness is of greater value than any gain.
It introduces us to the life of virtue
and the strength of our own holiness.

How few individuals, either rich or poor,
in this world feel they can afford
the space for Rightness.

If we become the potential to receive,
the GIVEN is accessible.

What Divine gifts of God to His Son
are forgiveness and gratefulness.
Without these one is blind.
This One Year Retreat is to help bring into focus
these beneficent Divine Forces
upon this planet.

Would you become a part of it?

The One Year Retreat is primarily for
the intensely earnest group of people
who dedicate their lives to:

I AM AMONG THE MINISTERS OF GOD.[4]

The Minister of God introduces his brothers to:

My only function is the one God gave me.[5]

Your coming to the One Year Retreat
is a statement of the life you have lived
and what you value,
for man's life is a statement of Divine Forces.

There is one life, and that I share with God.[6]

The curriculum is totally unambiguous, because the goal is not divided and the means and the end are in complete accord. You need offer only undivided attention. Everything else will be given you. For you really want to learn aright, and nothing can oppose the decision of God's Son. His learning is as unlimited as he is.[7]

Fulfillment

Unless man comes to fulfillment,
there can be no honesty.

What is fulfillment?
"It is a trust."
In what?
"In the perfection of WHAT IS."

Fulfillment is a trust
that becomes the blessed certainty.
There is no insecurity in it.
It makes one's life productive and universal.

Fulfillment is bigger than anything of the world.
Without fulfillment one is tormented.
For the Son of God,
ambition is not big enough.

To be precise is to relate only with Reality.

This is a course in how to know yourself.[8]

To this age of interpretation comes
the direct voice of *A Course In Miracles*
that brings us to the purity of One MIND,
the Mind of God.

Let me not see myself as limited.[9]

I am as God created me.[10]

My Self is holy beyond all the thoughts of holiness of which I now conceive. Its shimmering and perfect purity is far more

> *brilliant than is any light that I have ever looked upon. Its love is limitless, with an intensity that holds all things within it, in the calm of quiet certainty.*[11]

Everyone who has heard the call:

> *My only function is the one God gave me.*[12]

comes to action and is not delayed
by the abstract conflict of interpretations.
The means and the space are provided.
It is a matter of decision.
THE ACTION OF THE RETREAT ITSELF, BEING TOTAL, PROVIDES THE MEANS.

> *There is nothing my holiness cannot do.*[13]

THE ONLY ACTION OF THE WILL IS
THE DETERMINATION TO BE ONE WITH GOD.

> *. . . I have overcome the world.*
> John 16:33

is the direction of the retreat.

We are sharing the joy of our heart,
the joy of our calling;
we are not promoting.

A change of lifestyle is offered to us
free of charge.
You are to be the blessing of the world.

From my "Gratefulness Journal":

> "Fulfillment is masterful.
> It is not subject to a body.

I am not 'me.'
That is merely a habit.

I know God loves me.
I have no problems
and no suffering befalls a person
who calls nothing his own.

Religion is a state of fulfillment.
Externals do not touch it.
Divine Laws are not ruled by time."

The purpose of the Year Retreat is to rise to the State of *A Course In Miracles* and meet it.

You will get the means to get there.
We are sustained by the Love of God.

This day I will accept myself
as what my Father's Will created me to be.[14]

Let me remember that my goal is God.[15]

Be grateful.
Come with gladness.
The year is blessed by our need for the Peace of God.
You are surrounded by the Love of God.

My Self is ruler of the universe.[16]

Let every voice but God's
be still in me.[17]

I AM IN NEED OF NOTHING
BUT THE TRUTH.

"I Am in Need of Nothing but the Truth"

I sought for many things, and found despair. Now do I seek but one, for in that one is all I need, and only what I need. All that I sought before I needed not, and did not even want. My only need I did not recognize. But now I see that I need only truth. In that all needs are satisfied, all cravings end, all hopes are finally fulfilled and dreams are gone. Now have I everything that I could need. Now have I everything that I could want. And now at last I find myself at peace.

And for that peace, our Father,
we give thanks.
What we denied ourselves
You have restored,
and only that is what we really want.[18]

Part IX

The One Year Non-Commercialized Retreat

The One Year Non-Commercialized Retreat

The world is safe from love to everyone who thinks sin possible. Nor will it change. Yet is it possible what God created not should share the attributes of His creation, when it opposes it in every way?

If you could realize nothing is changeless but the Will of God, this course would not be difficult for you. For it is this that you do not believe. Yet there is nothing else you could believe, if you but looked at what it really is.[1]

Give your mind to the quiet purity of space
for wisdom requires relaxation.

We have put not only our heart into this letter
but much more.
Each word extends our very life.

For us this letter is an historical document
of man on the planet
during the twelfth hour of the twentieth century.
It is a spark of light in the midst
of the educated insanity of mankind.

From the very outset we had made clear
that the One Year Retreat was:

GOD'S PLAN FOR SALVATION

and not a personal project.

The Name of God cannot be sold,
thus it is non-commercialized.
It is a dedication to be consistent
with the Will of God.

The destiny of the New World is
to come to new consciousness
and to initiate this action of the Spirit
requires the wholeheartedness of:

My single purpose offers it to me.[2]

A Course In Miracles,
the most systematic, Divine order of education
ever known to man, is to help man
step out of the personality world
and be the God-created reality of his Being.

The Course is a gift of God to man
and a process of inner awakening.
It brings to our recognition the Mind of God
of which we are all a part.

Before embarking upon
the One Year Non-Commercialized Retreat
for the serious-minded,
I had spent the one year in silence myself
to see what the daily Lesson of
A Course In Miracles unfolds
and directly experience what transpires within,
given the space and the attention.

I had directly felt that within one hundred days
one could overcome insecurity
and the present-day obsession with survival.

How could I not share this with those
I have established a one-to-one relationship with?

Since *Love ye one another*
is the means of salvation and the ending of separation,
we had made the offer of a whole Year Retreat
for the first time in the history of the New World
free of all tuitions.

I wish you would see the purity and
the integrity of this action itself.
Knowing that such happenings and light
are upon this planet,
we gain strength by the Action of Grace upon us.

I, on my part, was determined not even to charge
for my boarding and lodging
but to pay for my own expenses
in order to be true to :

To give is to receive.[3]

We have continually emphasized the words of the Bible:

. . . SEEK YE FIRST THE KINGDOM OF GOD, . . .
AND ALL THINGS SHALL BE ADDED UNTO YOU.
Matthew 6:33

Thus we specified the qualifications and Laws necessary
to participate in the One Year Non-Commercialized Retreat.

Obviously, it has to be an action born
of your own energy and interest and
out of the order in your own life.

The disillusion with the external
was already to be in effect in you.
Also, your very self-centeredness was being
questioned — by you.

You had made your acquaintance with
A Course In Miracles and you were now prepared
to come to the One Year curriculum
of the Course to be awakened
by the daily lesson and to bring it to application.

Since the process of undoing was
already occurring in you,
the space to receive the Given was there.
The willingness to be consistent
with the Will of God had become important.
And you were to come to the retreat to be
amongst the Ministers of God.

I hope that the Year Retreat is
now put back into perspective.

We had stressed in letters and in *Holding Hands,*
the importance of relating with Eternal Laws
and that the purpose of the Foundation
was to be an extension of the Path of Virtue.

The very purpose of the Foundation is
the one-to-one relationship and
seeing a thing through to the end.
Thus you never were just a name to us
but a human being with whom we share
the One Life of God.

We offered a lasting relationship, friendship
and the joy of it,
valuing the commandment of:

Love ye one another.

We look upon each person
God brings us in contact with

as a lasting relationship.
This spirit continues irrespective of externals
or the Year Retreat.

As regarding the One Year Retreat,
at the Preliminary Retreat at Asilomar,
fifty-four people, fully knowing the costs and
the dedication involved, signed up for the Year,
above and beyond the one hundred and thirty-five
that had already applied for the Year.

Had it been a business venture,
we would have collected the deposit
right there and then.
But these were not the methods
with which one approaches what is holy
and is:

GOD'S PLAN FOR SALVATION.

We had spent over a year in the effort of research and finding a suitable place for this Holy Event. We had contacted every single, available retreat site in the country; approached chambers of commerce; made direct contact with ranches; people owning estates, campgrounds, and church retreats. We had placed ads in newspapers around the country; made, all told, over one thousand long distance phone calls; and taken ten plane trips to distant places to find the right location.

And in the later months, two people worked seven days a week, ten hours a day on this undertaking. A sum of $25,043 was spent on travel, telephone, postage, and printing. And this amount does not include any portion of salaries. At Asilomar, the committee of seven individuals of the One Year Retreat participants reviewed the accounts as well as the expenses the Foundation incurred on your behalf.

Then in our letter dated February 26, 1983,
when we had located selected places,
we informed you of the fifty participants
required to sign the lease
to be on time for Easter,
the day of Resurrection, April 3, 1983,
to start the Retreat.

But we have not received the money from
the fifty necessary participants;
thus the places we located could not be leased
even though we had already given notice to our landlord,
and those who work out to support themselves
had given notice to their employers.
It is alarming to see
that in the New World, with its abundance,
fifty people are not with the required purity of intent
to be with:

GOD'S PLAN FOR SALVATION.

You are among those who signed up at
the Preliminary Retreat to be a participant
or who sent an application for the Year.
Our motto, as you know, is:

"IN GOD WE TRUST"

and thus, in you, the human being, as well.
And we had on that trust dedicated ourselves
to finding a suitable place for the Year.

To those of you who were kind enough to send
a deposit or portion of it,
we are returning in a separate letter
the full amount of your check
without any deductions for expenses incurred.

Now we leave it to you,
because if we lose faith in each other's integrity,
then it is worse.
We need to be responsible.

We respect your sincerity and appreciate
your wanting to make it possible
for the One Year Retreat.
I pray for your growth and
for the relationship to continue between us.

What it says to each one of us is
that we have to come to some urgency and determination
to merit that which is of God.
The next step for each one is
not to put the Life of the Spirit in the second place
but to give authenticity to:

Salvation is my only function here.[4]

So we owe it to the Destiny of the New World
to come to:

"IN GOD WE TRUST"

for that is the vibration of this land.

We really should examine our lives and our priorities
and let us not value the valueless.
But wherever you are and whatever you are doing
I beseech you to stay close to
the daily lesson of *A Course In Miracles*
and to read the *Text* with a loving heart
before going to bed and thus bring a new purity
into the quality of your sleep.
Within a year's time,
it will transform your very consciousness.

It takes wisdom to come to relaxation,
and *A Course In Miracles* renews us daily
with its Divine gifts.
Please do find the time for quiet and
to transcend activity.

We would like to acquaint you with what a seer and a wise man had said about the One Year Retreat:

> "It is one of the golden opportunities upon this planet. But the question is: 'Who will be ready?' It will be a process of inner selection; a calling from within.
>
> At the Foundation for Life Action there will be, as there has been and is, an integrity that is a light that shines forth and draws people. The Foundation is like an open palm with no need to possess anything. The Foundation reminds. Its purpose and what it already is doing is to remind people of their relationship with God."

The times demand
that we listen to our heartbeats within,
for there is no truth in the external chaos.

Times will reveal themselves
that you cannot depend on the externals,
and without the externals there is no personality.

The time is coming when the fear of the external
will destroy the external.
It is a very painful thing,
because we identify with unreality.

There is no relationship at the personality level.
It is only the illusion of living

and not the extension of Life,
the holiness of Eternity.

We have to come to the life of ethics
with its order and discipline,
for we are of Heaven, subject only to Divine Laws
and not to manmade rules of punishment and reward.

I offer you my friendship.
Whether you come to the Year or not
does not affect my relationship with you.

Things at the time level are ever
the ways of change and endings.

Irrespective of whatever happens externally,
please know that you and I have a relationship
that the world of unreality cannot affect.
It is a flame that cannot be snuffed out.
You are the joy of my life.

This is not just a letter.
We want to get you acquainted with
the Mind of the Age and share with you
the urgency and the need for each one of us
to come to Divine Perspective and
a change of lifestyle.

We are not success or profit-oriented.
Our function is to be of service.
We do occasional workshops
and charge for the service we render
to pay for overhead.

We would meet the needs of the few
that have sold homes and are on the way,
for they have prepared and have come

to putting all alternatives away,
and we look forward to what form
the Year will take.

The retreat will begin in Los Angeles,
Easter, April 3, 1983.
A Course In Miracles is meant for people
who live in the world and do their daily work.
Those who are serious and want to come,
we will welcome and cooperate with.

It is the man who maintains
the truth and integrity of his word
who is given to know the Will of God.

Please continue to be part of what unfolds.
You are always welcome.
Come and be a part of what we are doing
and break bread with us.

Please know that the Christ is always with you
and never lose confidence in yourself.

For what else would Heaven help but
to bring one to God?
All the Universal Forces are there to help us
and in that we must trust.

The gift I received upon awakening today is:

> "I renew myself with the strength of Rightness.
> It makes me independent."

I value the one-to-one relationship with you
and extend my love
surrounding you with prayers daily.

Our Lesson today is:

My mind holds only what I think with God.[5]

No one can fail who seeks to reach the truth.[6]

I loose the world from all I thought it was.[7]

DELIVERANCE

Lead me, my Lord, to where my stillness is.
I seek my Father's Everlasting Arms
Which I alone can never hope to find,
For I am frail in seeking and in love.

Idols will come to hold their halting hands
Before me on my lonely journeying
Which will forever bar the way to me,
And I will faint in an illusion's grasp.

But come you with me and I cannot fail
To find my Father's house. As we approach
The holy gate, illusions shudder back
And angels come to offer us their wings.

I am the least and yet the greatest. I
Who hold your hand have Heaven's might with me.
I go in glory, for you walk with me.
Deliver me into my Father's Arms.[8]

Addenda

The Purpose of the Foundation for Life Action

The purpose of the Foundation for Life Action
is to be with the Eternal Laws
so that it does not become an organization.

LOVE IS ETERNAL.
ABILITIES EXTENDING LOVE ARE BLESSED.

In the absence of Love
abilities become the bondage of skills,
limited to personality.
Among virtuous men,
it is what the human being is that is Real,
and not what he does in a body.

The purpose of the Foundation is to be part of

GOD'S PLAN FOR SALVATION.[1]

Thus it has a different point of reference
than the thought system of man.

Obviously, the Name of God cannot be commercialized.
There are no fees in what we share.
We do not believe in loss and gain.
Non-commercialized action is provided by
the blessings of productive life.

"IN GOD WE TRUST."

Those who are with the Eternal Laws
in times of change remain unaffected.
In crisis, it is your care for another
that is your strength.

We have a function in the world
to be truly helpful to others,
knowing:

I am sustained by the Love of God.[2]

My only function is the one God gave me.[3]

Nothing real can be threatened.
Nothing unreal exists.[4]

We are not pressured by the brutality of success.
We are blessed by the work we do.
Gratefulness is complete, as love is independent.

To us, you, the human being, comes first.
Thus it enables us to go past
the conventional opinion of right and wrong
and relate directly to you.

For man is as God created him,
unchanged by the changeable society
that rules his body with its belief systems.

The Truth is a Fact that dissolves illusions of time.
Our function is to dispel the abstraction of ideas
and realize the actuality of Fact.

For,

I am under no laws but God's.[5]

The Purpose of
the Foundation for Life Action

Reverence for Life is of a still mind
hallowed by His Love.
This transformation is what we call

THE PATH OF VIRTUE.

The Path of Virtue is the ministry of gratefulness.

The wise who extends the Kingdom of God on earth
lives consistent with

> *"BUT SEEK YE FIRST THE KINGDOM OF GOD,*
> *AND HIS RIGHTEOUSNESS;*
> *AND ALL THINGS SHALL BE ADDED UNTO YOU."*
> Matthew 6:33

Biography of Tara Singh

Tara Singh is known as a teacher, author, poet, and humanitarian. The early years of his life were spent in a small village in Punjab, India. From this sheltered environment his family then traveled and lived in Europe and Central America. At twenty-two, his search for Truth led him to the Himalayas where he lived for four years as an ascetic. During this period he outgrew conventional religion. He discovered that a mind conditioned by religious or secular beliefs is always limited.

In his next phase of growth he responded to the poverty of India through participation in that country's postwar industrialization and international affairs. He became a close friend not only of Prime Minister Nehru and Mahatma Gandhi but also of Eleanor Roosevelt.

It was in the 1950's, as he outgrew his involvement with political and economic systems, that Mr. Singh was inpsired by his association with Mr. J. Krishnamurti and the teacher of the Dalai Lama. He discovered that mankind's problems cannot be solved externally. Subsequently, he became more and more removed from worldly affairs and devoted several years of his life to the study and practice of yoga. The discipline imparted through yoga helped make possible a three year period of silent retreat in Carmel, California, in the early 1970's.

As he emerged from the years of silence in 1976, he came into contact with *A Course in Miracles*. Its impact on him was profound. He recognized its unique contribution as a scripture and saw it as the answer to man's urgent need for direct contact withTruth. There followed a close relationship with its scribe. The Course has been the focal point of his life ever since.

Mr. Singh's love of the Course has inspired him to share it in workshops and retreats throughout the United States. He recognizes and presents the Course as Thoughts of God and correlates it with the great spiritual teachings and religions of the world.

From Easter 1983 to Easter 1984, Mr. Singh conducted the One Year Non-Commercialized Retreat: A Serious Study of *A Course in Miracles*. It was an unprecedented, in-depth exploration of the Course. No tuition was charged.

Mr. Singh continues to work closely with serious students under the sponsorship of the Foundation for Life Action, a school for bringing *A Course in Miracles* into application and for training teachers of *A Course in Miracles*. He is the author of numerous books and has been featured on many videotapes in which he discusses the action of bringing one's life into order, freeing oneself from past conditioning, and living the principles of the Course.

References

BOOK I

Preface

1. *A Course In Miracles* (ACIM) is a contemporary scripture which deals with the psychological/spiritual issues facing man today. It consists of three volumes: *Text* (I), *Workbook For Students* (II), and *Manual For Teachers* (III). The *Text* sets forth the concepts on which the thought system of the Course is based. The *Workbook For Students,* three hundred and sixty-five lessons, is designed to make possible the application of the concepts presented in the *Text.* The *Manual For Teachers* provides answers to some of the basic questions a student of the Course might ask and defines many of the terms used in the *Text.* (Editor)
2. ACIM, II, page 317.
3. Matthew 5:39.
4. The one commandment given by Jesus, "Love ye one another," appears many times in the New Testament. See, for example: John 13:34, 15:12, 15:17; Romans 13:8.
5. ACIM, I, Introduction.
6. ACIM, II, page 376.

BOOK II

Part I — *Gifts from the Retreat*

1. ACIM, I, Introduction.
2. ACIM, II, page 364.

Part II — *Holding Hands With You*

1. ACIM, II, page 177.
2. Ibid.
3. Ibid.
4. ACIM, I, page 109.

Part III — *The Preparation*

1. ACIM, II, page 223.
2. ACIM, II, page 432.
3. ACIM, II, page 186.

4. ACIM, II, page 213.
5. ACIM, II, page 406.

Part IV — *Excerpts from the Forty Days in the Wilderness*

1. ACIM, II, page 274.
2. ACIM, II, page 376.
3. ACIM, II, page 315.
4. ACIM, II, page 18.
5. ACIM, II, page 400.
6. ACIM, II, page 213.
7. ACIM, I, page 143.
8. ACIM, II, page 440.
9. Ibid.
10. ACIM, II, page 236.
11. ACIM, II, page 165.
12. ACIM, II, page 248.
13. ACIM, II, page 376.
14. ACIM, II, page 364.
15. ACIM, II, page 249.
16. ACIM, II, page 21.
17. ACIM, II, page 290.
18. ACIM, II, page 11.
19. ACIM, II, page 165.
20. ACIM, II, page 275.
21. ACIM, II, page 462.
22. ACIM, II, page 329.
23. ACIM, II, page 370.
24. ACIM, II, page 215.
25. ACIM, III, page 3.
26. ACIM, II, page 410.
27. ACIM, II, page 385.
28. ACIM, II, page 315.
29. ACIM, II, pages 376-377.
30. ACIM, II, page 132.
31. ACIM, II, page 274.
32. ACIM, II, page 329.
33. ACIM, II, page 462.
34. ACIM, II, page 317.
35. ACIM, II, page 462.
36. ACIM, II, page 476.
37. ACIM, II, page 290.
38. ACIM, II, page 18.

39. ACIM, II, page 132.
40. ACIM, II, page 165.
41. ACIM, II, page 376.
42. ACIM, II, page 469.
43. ACIM, II, page 390.
44. ACIM, I, page 109.

Part V — *Profile of Our Age*

1. ACIM, II, page 42.
2. ACIM, II, page 315.
3. ACIM, II, page 181.
4. ACIM, II, page 271.
5. ACIM, II, page 32.
6. ACIM, I, page 1.
7. ACIM, II, page 38.
8. ACIM, I, page 12.
9. ACIM, II, page 376.
10. ACIM, II, page 79.
11. ACIM, II, page 285.
12. ACIM, II, page 31.
13. ACIM, II, page 411.
14. ACIM, II, page 47.
15. ACIM, II, pages 43-44.
16. ACIM, II, page 45.
17. ACIM, II, page 397.
18. ACIM, II, page 274.
19. ACIM, II, page 315.
20. ACIM, I, page 1.

Part VI — *Non-Commercialized Life/The Path of Virtue*

1. See ACIM, I, pages 426-427; II, page 120 and following.
2. ACIM, II, page 191.
3. ACIM, II, page 189.
4. ACIM, II, page 58.
5. ACIM, II, page 195.
6. ACIM, II, page 132.
7. ACIM, II, page 216.
8. ACIM, II, page 132.
9. ACIM, II, page 120.
10. ACIM, I, page 51.
11. ACIM, I, page 48.

12. ACIM, I, page 49.
13. ACIM, I, page 50.
14. ACIM, I, page 56.
15. ACIM, I, page 55.
16. ACIM, I, page 67.
17. ACIM, II, page 119.
18. ACIM, I, page 54.
19. ACIM, II, page 65.
20. ACIM, II, page 177.
21. ACIM, II, page 189.
22. ACIM, II, page 53.
23. ACIM, II, page 141.
24. ACIM, II, page 130.
25. ACIM, II, page 150.
26. ACIM, II, page 132.
27. ACIM, II, page 151.
28. ACIM, II, page 104.
29. ACIM, II, page 188.
30. ACIM, II, page 174.
31. ACIM, II, page 177.
32. ACIM, I, page 84.
33. ACIM, I, page 24.
34. ACIM, II, page 159.
35. ACIM, II, page 192.
36. ACIM, II, page 164.
37. ACIM, II, page 166.
38. ACIM, II, page 168.
39. ACIM, II, page 203.
40. ACIM, II, page 172.
41. ACIM, II, page 203.
42. ACIM, II, page 190.
43. ACIM, I, page 128.
44. ACIM, I, Introduction.
45. ACIM, I, page 132.
46. ACIM, II, page 205.
47. ACIM, II, page 355.
48. ACIM, II, page 219.
49. ACIM, II, page 220.
50. ACIM, I, page 131.

Part VII — *Invitation to the One Year Non-Commercialized Retreat*

1. ACIM, II, page 235.
2. ACIM, I, page 152.
3. ACIM, II, page 218.
4. ACIM, I, page 151.
5. ACIM, II, page 195.
6. ACIM, II, page 308.
7. ACIM, II, pages 313-314.
8. Ibid.
9. ACIM, II, page 320.
10. ACIM, I, page 161.
11. ACIM, II, page 400.
12. ACIM, II, page 281.
13. ACIM, II, pages 347-348.
14. ACIM, II, page 119.
15. ACIM, II, page 321.
16. ACIM, II, page 321-322.
17. ACIM, I, page 159.
18. ACIM, II, page 259.
19. ACIM, II, page 310.
20. ACIM, II, page 308.
21. ACIM, II, page 162.
22. ACIM, II, page 308.
23. ACIM, II, page 281.
24. ACIM, II, page 177.
25. ACIM, II, page 342.

Part VIII — *"I Am in Need of Nothing but the Truth"*

1. ACIM, II, page 399.
2. ACIM, II, page 235.
3. ACIM, II, page 177.
4. ACIM, II, page 281.
5. ACIM, II, page 107.
6. ACIM, II, page 311.
7. ACIM, I, page 211.
8. ACIM, I, page 321.
9. ACIM, II, page 408.
10. ACIM, II, page 162.
11. ACIM, II, page 410.
12. ACIM, II, page 107.
13. ACIM, II, page 58.

14. ACIM, II, page 275.
15. ACIM, II, page 413.
16. ACIM, II, page 411.
17. Ibid.
18. ACIM, II, page 410.

Part IX — *The One Year Non-Commercialized Retreat*

1. ACIM, I, page 494.
2. ACIM, II, page 235.
3. ACIM, II, page 192.
4. ACIM, II, page 174.
5. ACIM, II, page 266.
6. ACIM, II, page 233.
7. ACIM, II, page 236.
8. *The Gifts of God,* page 55.

ADDENDA

The Purpose of the Foundation for Life Action

1. See ACIM, I, pages 426-427; II, page 120 and following.
2. ACIM, II, page 79.
3. ACIM, II, page 107.
4. ACIM, I, Introduction.
5. ACIM, II, page 132.

Other Materials by Tara Singh Related to A COURSE IN MIRACLES

BOOKS

Commentaries On A Course In Miracles
"Love Holds No Grievances" — The Ending Of Attack
A Course In Miracles — A Gift For All Mankind
The Future Of Mankind — The Branching Of The Road
How To Learn From A Course In Miracles
How To Raise A Child Of God (forthcoming)
Dialogues On A Course In Miracles (forthcoming)

AUDIO CASSETTE TAPES

A Course In Miracles Explorations, Series One: Origin, Purpose And Application
(three tape album with the book *How To Learn From A Course In Miracles)*
Bringing A Course In Miracles Into Application
(three tape album with the pamphlet *The School At "The Branching Of The Road")*
"What Is The Christ?" (three tape album with the book *A Course In Miracles — A Gift For All Mankind)*
The Heart Of Forgiveness (single tape)
Discussions On A Course In Miracles
(three tape album with the book *"Love Holds No Grievances" — The Ending Of Attack)* (forthcoming)
"What Is A Course In Miracles?" (two tape book pac) (forthcoming)
Tara Singh Tapes Of The One Year Non-Commercialized Retreat:
A Serious Study Of *A Course In Miracles*

VIDEO CASSETTE TAPES

"Nothing Real Can Be Threatened" —
A Workshop On A Course In Miracles
- Part I — *The Question And The Holy Instant*
- Part II — *The Deception Of Learning*
- Part III — *Transcending The Body Senses*
- Part IV — *Awakening To Self Knowledge*

Finding Your Inner Calling
How To Raise A Child Of God
Exploring A Course in Miracles (series)
- *What Is A Course In Miracles* and *"The Certain Are Perfectly Calm"*
- *God Does Not Judge* and *Healing Relationships*
- *Man's Contemporary Issues* and *Life Without Consequences*
- *Principles* and *Gratefulness*

A Call to Wisdom
and *A Call To Wisdom — Exploring A Course In Miracles*
Man's Struggle For Freedom From The Past
and *"Beyond This World There Is A World I Want"*
Life For Life and *Moneymaking Is Inconsistent With Life Forces*
The Call To Wisdom: A Discussion On A Course In Miracles
(Parts I & II)
"Quest Four" with Damien Simpson and Stacie Hunt
"Odyssey" and *"At One With"* with Keith Berwick

Additional copies of *The Voice That Precedes Thought* by Tara Singh may be obtained by sending a check, Mastercard or Visa number and expiration date to:

FOUNDATION FOR LIFE ACTION
902 South Burnside Avenue
Los Angeles, CA 90036
213/933-5591
Toll Free 1/800/732-5489
(Calif.) 1/800/367-2246

Limited edition, hardbound (plus $2.25 shipping/handling)	$21.95
Softcover (plus $1.75 shipping/handling)	$16.95

California residents please add 6 1/2% sales tax.

Thank you.